AA004952

Underwater Intervention (UI 2018)

New Orleans, Louisiana, USA
6-8 February 2018

ISBN: 978-1-5108-6247-0

Printed from e-media with permission by:

Curran Associates, Inc.
57 Morehouse Lane
Red Hook, NY 12571

Some format issues inherent in the e-media version may also appear in this print version.

Printed by Curran Associates, Inc. (2018)

For permission requests, please contact Underwater Intervention
at the address below.

Underwater Intervention
5206 Cypress Creek Parkway
Suite 202
Houston, Texas 77069
USA

Phone: 281-893-8539

www.underwaterintervention.com

Additional copies of this publication are available from:

Curran Associates, Inc.
57 Morehouse Lane
Red Hook, NY 12571 USA
Phone: 845-758-0400
Fax: 845-758-2633
Email: curran@proceedings.com
Web: www.proceedings.com

TABLE OF CONTENTS

COST SAVINGS USING MANNED UNDERWATER INTERVENTION IN SUITABLE WATER DEPTHS. EXPERIENCE FROM THE NORWEGIAN CONTINENTAL SHELF 1
R. Rossland

CITF OVERVIEW 19
D. McCarron

GATEWAY APPLICATIONS OF MARINE ROBOTICS IN THE OFFSHORE UNDERWATER ENERGY SECTOR 38
S. Pai

UNDERWATER EXAMINATIONS FOR CLASSIFICATION 44
J. Preston

BLUE FUTURES - ENGAGING THE NEXT GENERATION 51
L. Hotaling

INVENTION EDUCATION: FROM MATE ROV TEAM TO PATENT APPLICATION 59
N/A

DIVERS AUGMENTED VISION DISPLAY (DAVD) 79
D. Gallagher, R. Manley, A. Pilher, W. Hughes

LEVERAGING TECHNOLOGY IN LIVEBOATING 87
T. Detke

MATE COMPETITION UPDATE 99
J. Zande

UNMANNED MARITIME SYSTEMS FOR OCEANOGRAPHIC APPLICATIONS AND HYDROGRAPHIC MISSIONS 106
M. Bendzlowicz

RED TAIL DOWN: THE SEARCH FOR LAKE HURON'S MISSING TUSKEGEE AIRMEN 114
N/A

GAINING WORKPLACE EXPERIENCE THROUGH SEAMATE 158
D. Simpson, M. Diaz, N. Baryal, S. Charan

ALL-IN-ONE NAVIGATION: COMBINING DOPPLER AND INERTIAL TO IMPROVE ROV PERFORMANCE 166
I. Tena

STEM OUTREACH COPIAH-LINCOLN COMMUNITY COLLEGE 177
K. McKone, J. Crouse, C. Smith, D. Katt

DISCOVERYTM CT SCANNING: MITIGATING UNCERTAINTY IN PIPELINE INSPECTION AND ACCURATELY VISUALIZING AND CHARACTERIZING FLOW ASSURANCE ISSUES 186
J. Bramlett

WELCOME TO THE WORLD OF TRELLEBORG - SUBSEA BUOYANCY UNITS 202
W. Ricci

ASV SUPPORTED ROVS FOR IMR & MCM OPERATIONS 209
B. Anderson

WELCOME TO FOCAL TECHNOLOGIES - REDUCING ROV OPERATING COSTS AND ENHANCING RELIABILITY WITH MONITORING TECHNOLOGY 219
I. MacKay

SEAMATE ROV KITS PROGRESSION OF LEARNING 229
M. Gardner

DIVING & R.O.V. PICTURES 243
B. Merrill

HIGH LEVEL OVERVIEW OF NAVAL MINE COUNTERS MEASURES SUPPORT PROVIDED TO THE ROYAL CANADIAN NAVY AND DRDC-ATLANTIC 263
J. Covill, J. Podrebarac, T. Dunbar

GROWING A MARINE TECHNOLOGY PROGRAM FOR MIDDLE AND HIGH SCHOOL STUDENTS 281
R. Mcdonald, T. Miller-Way

MORRISON CONTRACTORS COMPANY OVERVIEW - SAFETY, ACCOUNTABILITY, QUALITY, SUSTAINABILITY IN OFFSHORE OPERATIONS 290
B. Ristic

INTEGRATED ROV SYSTEMS - HOW AND INTEGRATED STATE-OF-THE-ART ROV SYSTEM WAS USED IN THE DISCOVERY OF THE USS INDIANAPOLIS 298
M. Gilliam
CONNECTIVITY AND ROVS: GOING BEYOND THE DIVE 310
T. Curley
SUPERVISED AUTONOMY: HUMANS AND MACHINES 318
M. Ondler
ENGINEERING SUBSEA SYSTEMS USING REAL-TIME ROV/AUV SIMULATION 325
S. Dodd
Author Index

UNDERWATER INTERVENTION 2018

New Orleans, LA
February 6-8, 2018

CONFERENCE PROGRAM

PRESENTED BY ASSOCIATION OF DIVING CONTRACTORS INTERNATIONAL AND THE MARINE TECHNOLOGY SOCIETY'S ROV COMMITTEE

ADCI

CELEBRATING 50 YEARS 1968-2018

MARINE TECHNOLOGY SOCIETY ROV COMMITTEE

UI2018
UNDERWATER INTERVENTION

www.underwaterintervention.com

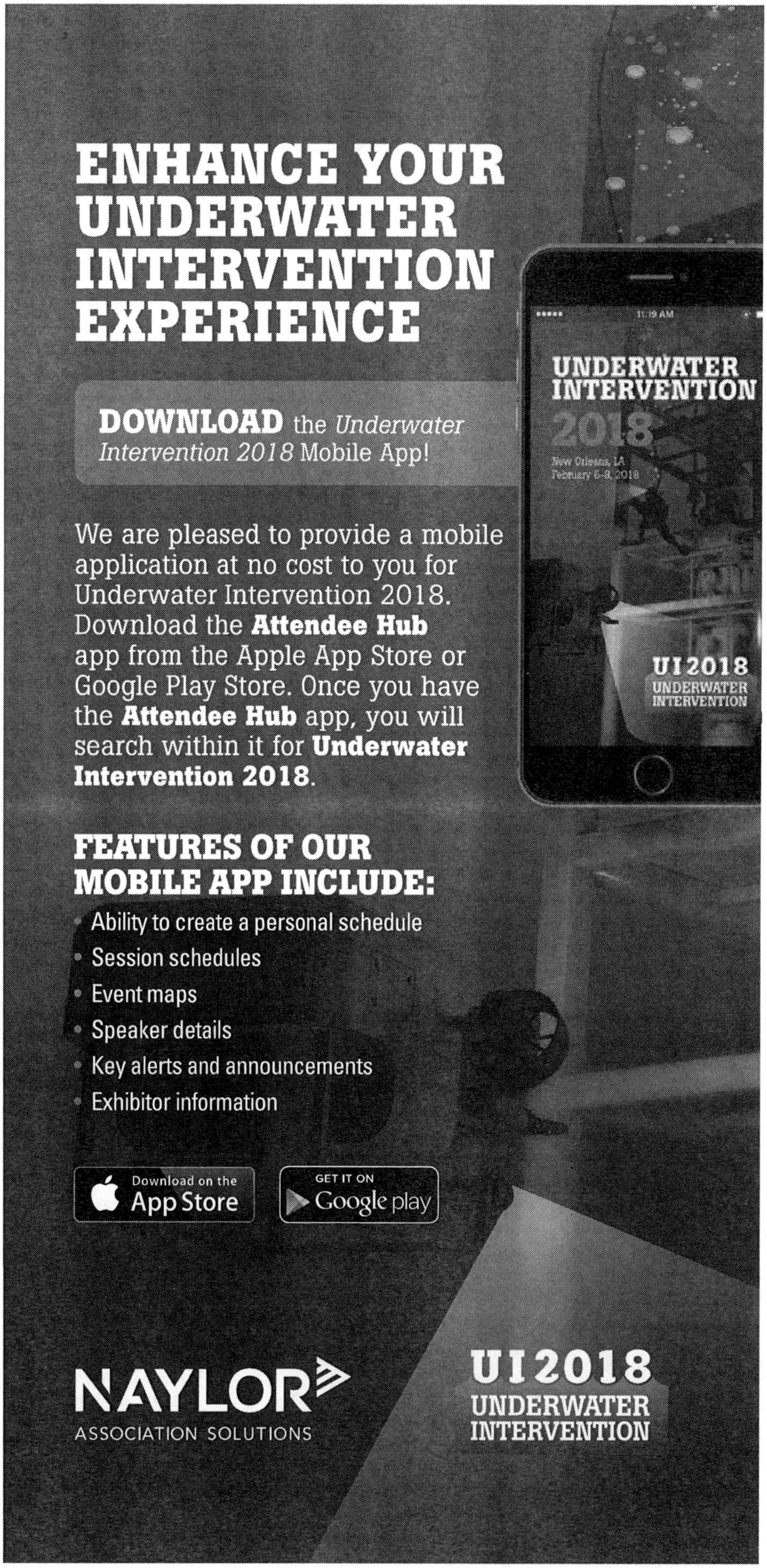
ENHANCE YOUR UNDERWATER INTERVENTION EXPERIENCE
DOWNLOAD the Underwater Intervention 2018 Mobile App!
We are pleased to provide a mobile application at no cost to you for Underwater Intervention 2018. Download the Attendee Hub app from the Apple App Store or Google Play Store. Once you have the Attendee Hub app, you will search within it for Underwater Intervention 2018.
FEATURES OF OUR MOBILE APP INCLUDE:
Ability to create a personal schedule
Session schedules
Event maps
Speaker details
Key alerts and announcements
Exhibitor information
Download on the App Store
GET IT ON Google play
UNDERWATER INTERVENTION
2018
New Orleans, LA
February 6-8, 2018
UI2018
UNDERWATER INTERVENTION
NAYLOR
ASSOCIATION SOLUTIONS
UI2018
UNDERWATER INTERVENTION

Leadership Committee

Conference Co-Chairmen

Chuck Richards, CA Richards and Assoc.,
MTS ROV Committee Chairman

Bryan Nicholls, U.S. Underwater Services, ADCI Board President

Conference Director

Phil Newsum, ADCI Executive Director

Technical Program Chairman

Steve Struble, Chapman Consulting, Inc.

Technical Program Track Leaders

Justin Manley, Just Innovation .... AUV & ASV

Steve Struble, Chapman Consulting, Inc. .... Commercial Diving

Mike Chapman, MECCO, Inc. .... Instruments and Sensors

Will Kohnen, Hydrospace Group .... Manned Submersibles

Jill Zande, MATE Center .... Marine Education/Academia

Bob Christ, Seatrepid .... ROV

Welcome to UI 2018

As Co-Chairmen of Underwater Intervention 2018, we are pleased to welcome you to New Orleans for three packed days of education, networking and fun! In today's competitive climate, it is our goal to provide as much value to all of our participants as possible. It is no secret that in order to flourish, we must evolve to best position ourselves for success. In that same spirit, we are evolving Underwater Intervention to best meet your needs and help position you and your company for success now and in the years to come. Of particular note this year is the newly added Underwater Technology Showcase, an area dedicated to notable advancements in technology, innovation, and development of the underwater operations industry.

As with past conferences, the Underwater Intervention Leadership Committee has been working tirelessly to deliver compelling, topical education. Organized in a thoughtful, well-planned schedule, you won't have to choose between sessions or time in the expo hall. There is dedicated time for both!

UI 2018 features available to all attendees include:

- UI Think Tank – Located in the hall for added convenience, we have planned several amazing sessions designed to be informative as well as interactive. Join any or all sessions to learn and share ideas on industry challenges and best practices.
- Underwater Technology Showcase – A must see addition to UI 2018, this feature encompasses groundbreaking feats of underwater innovation & technology! This year's showcase will feature The Blue Print Labs – join them for live demos and vibrant discussions that push the boundaries of underwater research and intervention.
- Supplier Solutions – Learn from our leading exhibitors who have their fingers on the pulse of our industry. See from their perspective the latest and greatest, innovative solutions for the industry challenges we face everyday.
- UI Café – no need to leave the convention center for a great New Orleans lunch...New Orleans is being served up in the exhibit hall just for you. A variety of New Orleans favorites with a selection of traditional comfort food will be hot and ready. Grab a bite and take advantage of the extra time you'll have in the expo hall!

It is our belief that Underwater Intervention is your single best platform to learn, share and network. It is designed specifically to be enriching and enjoyable. As we continue to make improvements, we welcome your feedback. In the meantime, we wish you all the best for success in all your underwater endeavors.

Best regards,

Bryan Nicholls
U.S. Underwater Services
ADCI President

Chuck Richards
C.A. Richards and Assoc.,
MTS ROV Chairman

Awards Dinner Keynote

Rear Admiral
Paul F. Thomas

This year's Awards Dinner will feature Keynote Speaker Rear Admiral Paul F. Thomas; Commander, 8th Coast Guard District.

He will discuss the state of the Outer Continental Shelf; the mission of the USCG in District 8; industry statistics; the challenges facing the offshore and underwater industry and the collaborative efforts to address them.

Headquartered in New Orleans, Admiral Thomas is responsible for Coast Guard operations spanning 26 states, including the Gulf of Mexico coastline from Florida to Mexico, the adjacent offshore waters and outer continental shelf, and the inland waterways of the Mississippi, Ohio, Missouri, Illinois and Tennessee River systems.

His previous Flag assignment was as the Assistant Commandant for Prevention Policy, where he was responsible for oversight and program management for all Coast Guard navigation services, boating safety, ports and facilities, waterways management, merchant mariner credentialing, vessel documentation, marine casualty investigation, commercial vessel inspections, and port state control.

A specialist in Marine Safety, Security and Environmental Protection, Admiral Thomas has served in countless capacities with the USCG.

Advisory Board

Don Walsh
PhD CAPT (Ret)

Jon Hazelbaker
Hammerhead Marine
Services, LLC

John Hughes
British Petroleum

Drew Michel
ROV Technologies Consulting

Dick Frisbie
Oceaneering International (ret)

Bev Morgan
Kirby Morgan Dive
Systems International

Early Bird Reception Sponsor

LOOKING TO THE FUTURE
BROCO 50 YEARS / 80 YEARS RANKIN
CELEBRATING PARTNERSHIPS

Partners

UNDERWATER

SEA TECHNOLOGY. Magazine

ON&T Ocean News & Technology www.oceannews.com

eco environment coastal & offshore

ROV PLANET

MARINE TECHNOLOGY REPORTER

DEEP TREKKER

BOATS & HARBORS
P.O. Box 647
Crossville, TN 38557
Phone: (931) 484-6100
Fax: (931) 456-2337

NAYLOR
ASSOCIATION SOLUTIONS

Special Events and Features

UI 2018 offers a variety of educational and networking opportunities beyond the technical sessions. Strategically developed to be informative and engaging our special features provide additional value for all attendees through more casual and interactive activities. After the technical sessions each day, hop down to the exhibit hall and join in on all the excitement while you also evaluate all the latest and greatest products and services for the underwater operations industry!

EARLY BIRD RECEPTION (SPONSORED BY BROCO, INC.)

Monday, Feb. 5, 6:00 PM – 8:00 PM,
La Nouvelle Orleans Ballroom Lobby, Convention Center (2nd Floor)

There is no better way to get a jump start on all the upcoming UI happenings than to get reacquainted with your colleagues in a fun and relaxing setting. Join us Monday for an evening filled with drinks, light food and live music.

UI THINK TANK

Located in the exhibit hall for added convenience, we have planned several amazing sessions designed to be informative as well as interactive. Join either or all sessions to learn and share ideas on industry challenges and best practices. This year's Think Tank features:

- Real-Time Simulation and Visualization for Underwater Operations; Presented by Felix Gorbatsevich, Pale Blue
- Latest NAVO Initiatives and General Charter; Presented by Commander Michael Bendzlowicz, U.S. Navy
- World War II Underwater Excavation; Presented by Ralph Wilbanks
- Investigation of the Effects of Oil Spills on the Environment and Public Health; Presented by GoMRI
- Stepping Towards Autonomy in Subsea Intervention; Presented by Paul Phillips, The Blueprint Lab

For more details and times, download the UI 2018 Mobile App

UI HAPPY HOUR IN EXHIBIT HALL

Tuesday, Feb. 6, 4:00 PM – 5:00 PM, Expo Hall

Wednesday, Feb. 7, 4:00 PM – 5:00 PM, Expo Hall

Thursday, Feb. 8, 2:00 PM – 3:00 PM, Expo Hall

As each day concludes, all are welcome to gather in the exhibit hall for Happy Hour.

Special Events and Features

UI SUPPLIER SOLUTIONS

Suppliers are an integral part of our industry. In order to provide the best products and services, they must have a superior understanding of industry challenges in order to resolve them. Sit in and learn from our leading exhibitors who have their fingers on the pulse of our industry. See from their perspective the latest and greatest innovations and best practices in developing solutions for the industry challenges we face every day. This year's Supplier Solutions feature:

- Don Kinney with Cylinder Training Services
- Dr. Harold (Bud) Vincent with Subsalve USA
- Rod Sanders with Cygnus Instruments, Inc.
- Arnauld Dumont with Dimensional Eye

For more details and times, download the UI 2018 Mobile App

UNDERWATER TECHNOLOGY SHOWCASE

The Underwater Technology Showcase is a must see addition to UI 2018. This feature encompasses groundbreaking feats of underwater innovation & technology! This year's showcase will feature The Blueprint Lab – join them for live demos and vibrant discussions that push the boundaries of underwater research and intervention.

HISTORICAL DIVING SOCIETY

The Historical Diving Society USA will be displaying surviving examples of the Morse and Schrader helmets, from the Lee Selisky Collection, that lead to the design of the U.S. Navy Mark V diving helmet. The Navy Mark V helmet saw service with the US military and other military services and commercial divers worldwide. The development of the Mark V design will be represented by five helmets on display at the HDS USA booth. The origins of the Mark V can be traced back into the 1800's, to the helmets that were retro actively designated USN Mark I to Mark IV. These designs influenced the final design of the Mark V which was introduced into service in 1916 and remained the standard US military helmet into the 1980's. It is believed that this is the first ever public display of all five USN Mark helmets.

UI CAFÉ

Tuesday-Thursday, Feb. 6-8, 12:00 PM – 1:00 PM in the Expo Hall

For added convenience, we are bringing a taste of New Orleans directly to you. No need to leave the convention center for a great New Orleans lunch, because we'll have a variety of New Orleans favorites with a selection of traditional comfort food hot and ready…right there in the exhibit hall. Grab a bite and take advantage of all the extra time you'll have in the expo hall!

Schedule At-A-Glance

For session descriptions, updated room locations, and more information, please refer to the Underwater Intervention 2018 Mobile App! Instructions on how to download can be found on page 2 of this guide.

Event rooms are subject to change. Please visit www.underwaterintervention, follow @UITradeShow on Twitter, or download the UI 2018 Mobile App for updates throughout the show!

TIME		LOCATION
Sunday, February 4th		
12:00 PM - 5:00 PM	Onsite Registration	*Exhibit Hall E*
1:00 PM - 5:00 PM	Exhibitor Move-In	*Exhibit Hall E*
3:00 PM - 5:00 PM	ADCI Executive Committee Meeting	*Marriott*
Monday, February 5th		
8:00 AM - 7:00 PM	Onsite Registration Open	*Exhibit Hall E*
8:00 AM - 5:00 PM	Exhibitor Move-In	*Exhibit Hall E*
8:00 AM - 4:00 PM	Tech Surge 1 & 2	*Rooms 354 & 352*
7:00 AM - 2:00 PM	ADCI Board Meeting	*Marriott*
2:00 PM - 4:00 PM	Technical Committee Meeting	*Room 355*
4:00 PM - 6:00 PM	ADCI Chapter Meetings	
6:00 PM - 8:00 PM	Early Bird Reception Sponsored by Broco, Inc.	Lobby - La Nouvelle Orleans Ballroom

Schedule At-A-Glance

For session descriptions, updated room locations, and more information, please refer to the Underwater Intervention 2018 Mobile App! Instructions on how to download can be found on page 2 of this guide.

TIME		LOCATION
Tuesday, February 6th		
8:00 AM - 5:00 PM	Onsite Registration	*Exhibit Hall E*
8:00 AM - 9:30 AM	ADCI Engineering Diving Committee	*Room 257*
8:30 AM - 10:30 AM	Concurrent Technical Sessions	*see track schedule*
10:30 AM - 10:45 AM	Break	
10:30 AM - 5:00 PM	Exhibit Hall Open	*Exhibit Hall E*
10:45 AM - 12:15 PM	Concurrent Technical Sessions	*see track schedule*
12:00 PM - 1:00 PM	Lunch in Exhibit Hall	*Exhibit Hall E*
12:30 PM - 5:00 PM	Concurrent Technical Sessions	*see track schedule*
1:00 PM - 2:00 PM	Think Tank	*Exhibit Hall E, Think Tank*
2:15 PM - 3:00 PM	Think Tank	*Exhibit Hall E, Think Tank*
3:00 PM - 3:30 PM	Supplier Solutions Session	*Exhibit Hall E, Supplier Solution*
3:45 PM - 4:30 PM	Think Tank	*Exhibit Hall E, Think Tank*
4:00 PM - 5:00 PM	Happy Hour	*Exhibit Hall E*
7:00 PM - 9:00 PM	Awards Dinner*	Marriott - River Bend Ballroom

Schedule At-A-Glance

For session descriptions, updated room locations, and more information, please refer to the Underwater Intervention 2018 Mobile App! Instructions on how to download can be found on page 2 of this guide.

TIME		LOCATION
Wednesday, February 7th		
8:00 AM - 5:00 PM	Onsite Registration	*Exhibit Hall E*
8:30 AM - 10:30 AM	Concurrent Technical Sessions	*see track schedule*
10:30 AM - 10:45 AM	Break	
10:30 AM - 5:00 PM	Exhibit Hall Open	*Exhibit Hall E*
10:45 AM - 12:15 PM	Concurrent Technical Sessions	*see track schedule*
11:45 AM - 1:30 PM	ADCI Marketing Committee Lunch	*Room 257*
12:00 PM - 1:00 PM	Lunch in the Exhibit Hall	*Exhibit Hall E*
2:00 PM - 3:00 PM	Think Tank	*Exhibit Hall E, Think Tank*
3:00 PM - 3:45 PM	Supplier Solutions Session	*Exhibit Hall E, Supplier Solution*
3:45 PM - 4:15 PM	Supplier Solutions Session	*Exhibit Hall E, Supplier Solution*
4:00 PM - 5:00 PM	Happy Hour	*Exhibit Hall E*
5:00 PM - 5:30 PM	ADCI General Membership Meeting	*Room 252*
5:30 PM - 7:00 PM	ADCI Board Meeting	*Room 257*

Schedule At-A-Glance

For session descriptions, updated room locations, and more information, please refer to the Underwater Intervention 2018 Mobile App! Instructions on how to download can be found on page 2 of this guide.

TIME		LOCATION
Thursday, February 8th		
8:00 AM - 1:00 PM	Onsite Registration	*Exhibit Hall E*
8:00 AM - 5:00 PM	IOGP - DIWG Meeting	*Room 254*
8:30 AM - 10:30 AM	Concurrent Technical Sessions	*see track schedule*
10:00 AM - 12:00 PM	ROV Committee Meeting	*Room 257*
10:30 AM - 10:45 AM	Break	
10:30 AM - 3:00 PM	Exhibit Hall Open	*Exhibit Hall E*
10:45 AM - 12:15 PM	Concurrent Technical Sessions	*see track schedule*
12:00 PM - 1:00 PM	Lunch in the Exhibit Hall	*Exhibit Hall E*
1:30 PM - 2:00 PM	Think Tank	*Exhibit Hall E, Think Tank*
2:00 PM - 2:30 PM	Supplier Solutions Session	*Exhibit Hall E, Supplier Solution*
2:00 PM - 3:00 PM	Happy Hour	*Exhibit Hall E*
3:00 PM - 8:00 PM	Exhibitor Move-Out	*Exhibit Hall E*
Friday, February 9th		
8:00 AM - 12:00 PM	Exhibitor Move-Out	*Exhibit Hall E*

Schedule by Technical Track

For session descriptions, updated room locations, and more information, please refer to the Underwater Intervention 2018 Mobile App! Instructions on how to download can be found on page 2 of this guide.

AUV & ASV TRACK - ROOM 355

Tuesday - Feb. 06, 2018

TIME	PRESENTATION TITLE	SPEAKER
Session Title: Unmanned Maritime Vehicles I		
09:00 - 09:30	***Unmanned Maritime Vehicles (UMVs) for Offshore Operations: Decoding the Alphabet Soup***	Justin Manley, Just Innovation
09:30 - 10:00	***Teledyne marine family of unmanned systems: Recent operational results***	Arnar Steingrimsson, Teledyne Marine
10:00 - 10:30	***SeaScout Autonomous Unmanned Underwater Vehicle (UUV)***	James E. (Jim) Hardiman, QinetiQ North America
10:30 - 10:45	**BREAK**	
Session Title: Unmanned Maritime Vehicles II		
10:45 - 11:15	***The Saab Underwater Robotics Platform for Hybrid AUV/ROV systems***	Chris Roper, SAAB
11:15 - 11:45	***New concepts of operations enabled by compact affordable UUVs***	Bill Key, Riptide Autonomous Solutions
11:45 - 12:15	***Subsea Communications for AUV and USV collaboration***	Ioseba Tena, Sonaryne

Wednesday - Feb. 07, 2018

TIME	PRESENTATION TITLE	SPEAKER
Session Title: Unmanned Maritime Vehicles III		
08:30 - 09:00	***Autonomous Surface Vessels for Offshore Operations***	Larry Karl, AVS Global
09:00 - 09:30	***Wave Energy USVs Supporting Oil and Gas Industry***	Sean Halpin, Liquid Robotics
09:30 - 10:00	***Use of autonomous underwater gliders for geochemical exploration surveys***	Walt McCall, Blue Ocean Monitoring

Schedule by Technical Track

For session descriptions, updated room locations, and more information, please refer to the Underwater Intervention 2018 Mobile App! Instructions on how to download can be found on page 2 of this guide.

AUV & ASV TRACK - ROOM 355

TIME	PRESENTATION TITLE	SPEAKER
10:00 - 10:30	***Glider Operations with Advanced Sensors***	Chad Lemke, University of South Florida
10:30 - 10:45	**BREAK**	
	Session Title: Unmanned Maritime Vehicles IV	
10:45 - 11:15	***Unique Delivery System - Part II***	Jennifer White, Ocean Infinity
11:15 - 11:45	***Buoyancy: A crucial component of vehicle design***	Ruth Clay, Trelleborg Offshore
11:45 - 12:15	***The emerging role of autonomy in modern combat swimmer propulsion systems***	Ben Kinnaman, Greansea Systems

Schedule by Technical Track

For session descriptions, updated room locations, and more information, please refer to the Underwater Intervention 2018 Mobile App! Instructions on how to download can be found on page 2 of this guide.

MARINE EDUCATION
ACADEMIA TRACK - ROOM 354

Tuesday - Feb. 06, 2018

TIME	PRESENTATION TITLE	SPEAKER
Session Title: History, Preservation PLUS Student Presentations		
09:00 - 09:30	***Red Tail Down***	Stockbridge Community Schools
09:30 - 10:00	***Invention Education***	Bob Richards
10:30 - 10:45	**BREAK**	

Wednesday - Feb. 07, 2018

TIME	PRESENTATION TITLE	SPEAKER
Session TItle: Successful Tools for Technology Education PLUS Student Presentations		
08:30 - 09:00	***MATE Center's progression of learning:***	Matt Gardner, MATE Center
09:00 - 09:30	***Gaining workplace experience through SeaMATE***	SeaMATE Community College Students
09:30 - 10:00	***MATE international ROV competition***	Jill Zande, MATE Center
10:00 - 10:30	***History of Diving and ROV's and impact on the oilfield***	Bryce Merrill, idea works
10:30 - 10:45	**BREAK**	
Session Title: Successful Tools for Technology Education		
10:45 - 11:15	***Blue Futures: Engaging the next generation***	Liesl Hotaling, Eidos Education
11:15 - 11:45	***Growing a marine technology program for middle and high school students***	Tina Miller-Way
11:45 - 12:15	***ROV Outreach***	Kevin McKone, Copiah-Lincoln Community College

For session descriptions, updated room locations, and more information, please refer to the Underwater Intervention 2018 Mobile App! Instructions on how to download can be found on page 2 of this guide.

MANNED SUBMERSIBLES TRACK - ROOM 357

Tuesday - Feb. 06, 2018

TIME	PRESENTATION TITLE	SPEAKER
	Session Title: Ocean Exploration 1	
08:30 - 09:00	***20:17-20:18 MUV Industry Overview***	William Kohnen, Hydrospace Group
09:00 - 09:30	***2017 Alvin Operations and 6500 meter conversion overview***	Anthony Tarantino, Alvin Group, WHOI
09:30 - 10:00	***Shinkai 6500 - On operation after remodeling inside the pressure hull***	Masanobu Yanagitani, JAMSTEC Takuma Onishi, NME
10:00 - 10:30	***Hazards of in-hull lithium ion batteries and hazard mitigation in DSV Alvin***	W. Bruce Strickrott, Alvin Group, WHOI
10:30 - 10:45	**BREAK**	
	Session Title: Ocean Exploration 2	
10:45 - 11:15	***Development of Manned Submersible 'MATSYA 6000'***	Dr. G.A. Ramadass, National Institute of Ocean Technology
11:15 - 12:15	***30 years Anniversary of MIR submersibles creation***	Dr. Anatoly Sagalevich, PP Shirshov Inst Oceanology RAS

Schedule by Technical Track

For session descriptions, updated room locations, and more information, please refer to the Underwater Intervention 2018 Mobile App! Instructions on how to download can be found on page 2 of this guide.

MANNED SUBMERSIBLES TRACK - ROOM 357

TIME	PRESENTATION TITLE	SPEAKER
01:00 - 01:30	***RIDE Year in Review 2017***	Karl Stanley, Roatan Inst of Deepsea Exploration
01:30 - 02:00	***Building a New Generation of Composite Submersibles for Tourism and Recreation***	Adam Wright, DEEPFLIGHT
02:00 - 02:30	***2017 ICTINEU 3 exploration of Nice canyon to 1000m depth***	Carme Parareda, ICTINEU Submarins S.L.
03:30 - 04:00	***Blue Tech: Driving Opportunity through Policy, Outreach and Investment***	Collen Hahn, Gryphon Media Strategies
04:00 - 04:30	***Blue Tech: Driving Opportunity through Policy, Outreach and Investment***	Collen Hahn, Gryphon Media Strategies
Wednesday - Feb. 07, 2018		
Session Title: Submersible Operations - PART 1		
08:30 - 09:00	***United States Navy Submarine Rescue Systems Deployment to Argentina, Commercial and Industry Interaction***	Capt. Mark Hazenberg, US Navy
09:00 - 09:30	***Pisces VI Deep Sea Submarine Renovation***	Scott Waters, Pisces VI Deep Sea Submarine
09:30 - 10:00	***ICTINEU 3 – 2016 Archaeological Expedition in the Mediterranean and long range capacity confirmed***	Pere Fores, ICTINEU Submarins S.L.

Schedule by Technical Track

For session descriptions, updated room locations, and more information, please refer to the Underwater Intervention 2018 Mobile App! Instructions on how to download can be found on page 2 of this guide.

MANNED SUBMERSIBLES TRACK - ROOM 357

TIME	PRESENTATION TITLE	SPEAKER
10:00 - 10:30	***SEAmagine's HOV Pilot Training in 2017 for Argentinian Coast Guard***	Charles Kohnen, SEAmagine Hydrospace Corp
10:30 - 10:45	**BREAK**	
	Session Title: Submersible Operations - PART 2	
10:45 - 11:15	***History of Oceaneering Deepest Pipeline Repair at 2120 ft with WASP ADS SUIT***	Bob Frangenberg, Oceaneering International
11:15 - 11:45	***New Manned Submersible Solutions Democratize Access to the Oceans***	Graham Hawkes, HAWX Open Ocean
11:45 - 12:15	***MUV Operations Consensus Standard - Concept Overview***	William Kohnen, MTS MUV Committee Kip Peterson, Thorsborg Institute
12:15 - 01:00	**LUNCH**	
01:00 - 01:30	***WORKSHOP - MUV Operations Consensus Standard***	Kip Peterson, MTS MUV Committee
01:30 - 02:00	***WORKSHOP - MUV Operations Consensus Standard***	Kip Peterson, MTS MUV Committee
01:00 - 03:00	***MTS MUV Annual Committee Meeting***	William Kohnen, MTS MUV Committee

Schedule by Technical Track

For session descriptions, updated room locations, and more information, please refer to the Underwater Intervention 2018 Mobile App! Instructions on how to download can be found on page 2 of this guide.

MANNED SUBMERSIBLES TRACK - ROOM 357

Thursday - Feb. 08, 2018

TIME	PRESENTATION TITLE	SPEAKER
	Session Title: Submersible Technology	
09:00 - 09:30	***Near VR presence for manned/ unmanned Applications***	Leonard Pool, Sidus Solutions
09:30 - 10:00	***Honeycomb Window Seat Non-Linear FEA Study & Testing***	Thomas Trudell, Roatan Inst of Deepsea Exploration,
10:00 - 10:30	***Significance of Heat Retention to Structural Integrity of Glassy Polymer Viewports***	Bart Kemper PE, Linda Cross Kemper Engineering Services

Schedule by Technical Track

For session descriptions, updated room locations, and more information, please refer to the Underwater Intervention 2018 Mobile App! Instructions on how to download can be found on page 2 of this guide.

COMMERCIAL DIVING TRACK - ROOM 352

Tuesday - Feb 06, 2018

TIME	PRESENTATION TITLE	SPEAKER
09:30 - 10:00	***Initial Takeaways from the ADCI Audit Initiative***	Phil Newsum, ADCI
10:00 - 10:30	***Initial Takeaways from the ADCI Audit Initiative***	Jon Hazelbaker
10:30 - 10:45	**BREAK**	
10:45 - 11:15	***Improving Labor Relations Through Training***	Guadalupe Vargas Jr., Carpenters International Training Fund
11:15 - 11:45	***The inspection and rehabilitation of a three-quarter century old pier***	Andrew Fulton, Pennoni
11:45 - 12:15	***Case Study of underwater coatings (NASA NBL Pool)***	Brett Crawford

Wednesday - Feb. 07, 2018

TIME	PRESENTATION TITLE	SPEAKER
09:00 - 09:30	***Divers Augmented Vision Display (DAVD) System***	Dennis Gallagher, Naval Surface Warfare Center-Panama City Division
09:30 - 10:00	***Divers Umbilicals***	Erich Vargas, Umbilicals International
10:00 - 10:30	***Diving, as intervention method***	Rolf Røssland, NUI
10:30 - 10:45	**BREAK**	

Schedule by Technical Track

For session descriptions, updated room locations, and more information, please refer to the Underwater Intervention 2018 Mobile App! Instructions on how to download can be found on page 2 of this guide.

COMMERCIAL DIVING TRACK - ROOM 352

TIME	PRESENTATION TITLE	SPEAKER
10:45 - 11:15	***Hyperbaric Medicine Update***	Dr. Joe Serio
11:15 - 11:45	***Hyperbaric Medicine Update***	Dr. Tony Alleman
11:45 - 12:15	***Hyperbaric Medicine Update***	Dr. Robert Bourgeois
	Thursday - Feb. 08, 2018	
09:00 - 09:30	***Nitrox Diving from a Daughter Craft - A Case Study***	Mark Lonsdale, Oceaneering International
09:30 - 10:00	***Leveraging Technology in Liveboating***	Travis Detke, Aqueos Corporation
10:00 - 10:30	***Safety, Accountability, Quality, Sustainability in Offshore Operations***	Bo Ristic, Chet Morrison Contractor
10:30 - 10:45	**BREAK**	
10:45 - 11:15	***Hazards that have been overlooked in U/W Burning***	John Roat, Greg Cain
11:15 - 11:45	***Oxy Arc Burning Safety Training & Certification***	John Roat, JCRoat Subject Matter Expert Services LLC Greg Cain

Schedule by Technical Track

For session descriptions, updated room locations, and more information, please refer to the Underwater Intervention 2018 Mobile App! Instructions on how to download can be found on page 2 of this guide.

INSTRUMENTS & SENSORS/ OCEAN RESEARCH & MARINE ARCH TRACK - ROOM 254

Tuesday - Feb. 06, 2018

TIME	PRESENTATION TITLE	SPEAKER
11:15 - 11:45	***Liftbags Proper Selection and Maintenance of, with Repair Tips***	Richard G. Fryburg

Wednesday - Feb. 07, 2018

TIME	PRESENTATION TITLE	SPEAKER
08:30 - 09:00	***Compact solutions for tracking divers***	Ioseba Tena, Sonadyne
09:00 - 09:30	***Universal Underwater Attachment Method***	Robert Conner
09:30 - 10:00	***Marine Learning-Based Systems: New Intelligent Device to Enhance Underwater Navigation, Positioning, and Vehicle Control Systems for Multipurpose Applications***	Yulian Mahendra, IKM Subsea
10:00 - 10:30	***Near VR presence for manned/ unmanned Applications***	Leonard Pool, Sidus Solutions
10:30 - 10:45	**BREAK**	
10:45 - 11:15	***Research Updates from the Pisces VI Submarine***	Grace Young
11:15 - 11:45	***ICTINEU 3 – 2016 Archaeological Expedition in the Mediterranean***	Carme Parareda, ICTINEU Submarins S.L.
11:45 - 12:15	***Overview of the Soft Biomimetic Underwater Surveillance Vehicle Mathsyais***	Siddharth Sridhar

For session descriptions, updated room locations, and more information, please refer to the Underwater Intervention 2018 Mobile App! Instructions on how to download can be found on page 2 of this guide.

ROV TRACK - ROOM 356

Tuesday - Feb. 06, 2018

TIME	PRESENTATION TITLE	SPEAKER
09:00 - 09:30	***Discovery™ Computed Tomography (CT) scanning: mitigating uncertainty in pipeline inspection by accurately visualising and characterising flow assurance issues***	Jim Bramlett, Tracerco
09:30 - 10:00	***Schlumberger Robotics Case Studies***	Sudhir Pai, Schlumberger Robotics Services
10:00 - 10:30	***Supervised Autonomy: Humans AND Robots***	Matt Ondler, Houston Mechatronics Inc
10:30 - 10:45	**BREAK**	
10:45 - 11:15	***The Technology Direction and Challenges of Underwater Surveillance Methodology within Canada's Naval Mine Counter Measures Community***	Joseph Podrebarac, Canadian Department of National Defence
11:15 - 11:45	***Ultimate Visibility Tethers***	Alex Parr, Lumiflex Corporation
11:45 - 12:15	***Vision based technique of position and orientation estimation***	Hyeonseung Ki, Korea Institute of Robot and Convergence

Schedule by Technical Track

For session descriptions, updated room locations, and more information, please refer to the Underwater Intervention 2018 Mobile App! Instructions on how to download can be found on page 2 of this guide.

ROV TRACK - ROOM 356

Wednesday - Feb. 07, 2018

TIME	PRESENTATION TITLE	SPEAKER
09:00 - 09:30	***UAVs in Offshore Inspection Operations***	John Ferguson
09:30 - 10:00	***Buoyancy: A crucial component of vehicle design***	Ruth Clay, Trelleborg Offshore
10:00 - 10:30	***ABS UWILD Rules and Guidelines***	John Preston, American Bureau of Shipping
10:30 - 10:45	**BREAK**	
10:45 - 11:15	***Reducing ROV Operating Costs and Enhancing Reliability with Monitoring Technology***	Shawn Taylor
11:45 - 12:15	***Integrated State-of-the-Art ROV System used in the Discovery of the USS Indianapolis***	Marybeth Gilliam

Schedule by Technical Track

For session descriptions, updated room locations, and more information, please refer to the Underwater Intervention 2018 Mobile App! Instructions on how to download can be found on page 2 of this guide.

ROV TRACK - ROOM 356

Thursday - Feb. 08, 2018

TIME	PRESENTATION TITLE	SPEAKER
09:00 - 09:30	***Connectivity and Remotely Operated Vehicles (ROVs)***	Ted Curley
09:30 - 10:00	***Engineering Subsea Systems Using Real-Time ROV/AUV Simulation***	Steve Dodd
10:00 - 10:30	***All-in-one Navigation: Combining Doppler and Inertial to improve ROV performance***	Ioseba Tena
10:30 - 10:45	**BREAK**	
11:15 - 11:45	***USV/ROV integrations for MCM Operations***	Brian Anderson
11:45 - 12:15	***Update on the Eelume: A Flexible and Subsea Resident IMR vehicle***	Arnt Olsen

Exhibitor Directory

A

AK Industries.................................. **427**
3115 E. Las Hermanas Street
Rancho Dominguez, CA 90221
310-762-1600 • F: 310-762-1616
sales@ak-ind.com
www.ak-ind.com

AK Industries is a agile, high-tech manufacturer of underwater electrical connectors specializing in building the most rugged interconnect products.

Anekonnect Inc............................. **545**
19849 Nordhoff Street
Northridge, CA 91324
818-217-5008
contact@anekonnect.io
www.anekonnect.io

Anekonnect is an online Cable Design Platform supporting users customize their design of underwater cables and connectors.

Aqua Lung America...................... **626**
2340 Cousteau Court
Vista, CA 92081
800-775-3483
sreilly@aqualung.com
www.aqualung.com/militaryandprofessional

Aqua Lung supports Military & Professional divers worldwide. Your One Stop™ for Drywear, Helmets, Rebreathers, Regs, BCs, Harnesses, and accessories.

Aqua Tech Solutions.................... **321**
115 Nova Drive
Broussard, LA 70518
337-837-3999 • F: 337-837-3344
jennifer.finley@aqua-techsolutions.com
www.aqua-techsolutions.com

Aqua-Tech Solutions sets the industry standard for excellence in reliable underwater electronic and mechanical equipment rental, sales, and service.

Aqua-Air Industries, Inc.............. **521**
639 Manhattan Boulevard
Harvey, LA 70058
504-362-8124 • F: 504-362-3600
sales@aquaairind.com
www.aquaairind.com

45 years of experience in manufacturing and supplying 100s of commercial diving products including shallow diving products and saturation systems.

Armada Systems, Inc................... **441**
508 Mountain Drive
Destin, FL 32541
850-664-5197 • F: 850-243-4009
armadahull@aol.com
www.armadahull.com

Armada Systems mfg underwater cleaning equipment. Equipment is used for cleaning ship hulls, propellers, and all underwater structures.

Avon Protection Systems........... **436**
1361 Brass Mill Road, Suite F
Belcamp, MD 21017
410-273-1161
mollai.dolan@avon-protection.com
www.avon-protection.com

Avon Protection provides military diving life support equipment and tactical products.

Exhibitor Directory

AXSUB Inc 737
112 Montee Industrial #200
Rimouski, QC G5M 1B1
Canada
418-731-1539 • F: 418-731-1557
sales@axsub.com
www.axsub.com

AXSUB is world leader in Dive Data Management Systems (DMS) for commercial diving industry that are compliant with the new NORSOK standard.

B

Bay-Tech Industries, Inc. 443
1253 First Avenue
Harvey, LA 70058
504-328-6456 • F: 504-328-6458
sales@baytechrentals.com
www.baytechrentals.com

Bay-Tech is a commercial diving equipment supplier founded in 1995 to provide quality equipment & supplies to the diving industry around the world.

Beta Subsea Systems 318
19210 Cohen Green
Houston, TX 77094
713-419-8683
bbaugh@baughengrs.com
www.betasubsea.com

Beta Subsea Systems, Div. Baugh Consulting Engineers, Inc. Development, manufacture, and licensing of subsea and pipeline IP.

BIRNS, Inc. 720
1720 Fiske Place
Oxnard, CA 93033
805-487-5393 • F: 805-487-0427
service@birns.com
www.birns.com

BIRNS, Inc. is an ISO 9001:2008 and NAVSEA PRO-020 certified global leader in high performance connectors, cable assemblies and lighting systems.

Blue Robotics, Inc. 640
4030 Spencer Street, Suite 102
Torrance, CA 90503
609-675-5405
elisa@bluerobotics.com
bluerobotics.com

Blue Robotics provides high quality, low cost components for marine robotics.

Boats & Harbors 742
175 4th Street
Crossville, TN 38572
931-484-6100
gdonnelly@boats-and-harbors.com
www.boats-and-harbors.com

Meeting Commercial Divers Equipment Needs Since 1958!

Broco Inc. 723
400 S. Rockefeller Avenue
Ontario, CA 91761
909-483-3222 • F: 909-483-3233
gjones@brocoinc.com
www.brocoinc.com

Broco® Underwater's patented products and proprietary formulas make underwater cutting and welding jobs safer, faster, and more economical.

C

CaviDyne, LLC 247
5077 Fruitville Rd, Suite 109-157
Sarasota FL 34232
352-275-5319
sales@cavidyne.com
www.cavidyne.com

Cochran Consulting, Inc.............. 737
1758 Firman Drive
Richardson, TX 75081
972-644-6284
salesmanager@DiveCochran.com
www.CochranMilitary.com

Cochran Undersea Technology & Cochran Military designs, manufactures, and markets state-of-the-art diving computers made in Richardson, TX U.S.A.

Copenhagen Subsea A/S............ 243
Trolijevej 2
Hellerup, Denmark 2900
www.copenhagensubsea.com
+45 70221162

Product Categories: Propulsion Systems & Thrusters - ROV, AUV

Cygnus Instruments, Inc.............. 736
6900 Phillips Highway, Suite 8
Jacksonville, FL 32216
410-267-9771
sales@cygnusinstruments.com
www.cygnusinstruments.com

Cylinder Training Services......... 732
PO Box 531
Edmonds, WA 98020
206-601-8993
info@cylindertrainingservices.com
www.cylindertrainingservices.com

CTS (ISO compliant) provides training for users of HP tanks. It meets OSHA, NFPA and CGA standards with $2 Million errors and omissions insurance.

Exhibitor Directory

D

Darkwater Vision........................... 737
1710 N. Hercules Avenue, Suite 103
Clearwater, FL 33765
727-400-4145 • F: 727-499-7117
sales@darkwatervision.com
www.darkwatervision.com

Our Infrared based vision system allows divers to see through tannin stained and silted waters without having to displace water. A true game changer!

David Clark Company.................. 637
360 Franklin Street
Worcester, MA 01604
800-900-3434 • F: 508-751-5927
bdaigle@davidclark.com
www.davidclark.com

David Clark has been dedicated since 1935 to design, development and manufacture of products that protect people and save lives.

Deep Ocean Engineering, Inc. ...341
2403 Qume Drive
San Jose, CA 95131
408-436-1102 • F: 408-436-1108
cocolai@deepocean.com
www.deepocean.com

Deep Ocean Engineering is a tech company in the underwater robotics business. We design and build the highest quality ROVs and USVs in the world.

DeepSea Power & Light, Inc. 542
4033 Ruffin Road
San Diego, CA 92123
858-576-1261
sales@deepsea.com

DeepSea Power & Light has over 30 years of experience manufacturing underwater lights, cameras, batteries, and lasers. Learn more at www.deepsea.com.

Denso North America.................. 527
9747 Whithorn Drive
Houston, TX 77095
281-821-3355
info@densona.com
www.densona.com

Denso North America is a leading manufacture of corrosion protection products to provide protection to new piles and provide pile rehabilitation.

Digital Edge Subsea..................... 644
Doubletree Court
10 Cavendish Street
Ulverston
United Kingdom
+44-1229-206456
john@digitaledgesubsea.com
www.digitaledgesubsea.com

Digital Edge Subsea develops and manufactures DVR products for use throughout the offshore oil & gas industry including diving, ROV and topside ops.

Dimensional Eye........................... 335
5115 Douglas Fir Road Unit B
Calabasas, CA 91302
310-699-6227
arnauld@dimeye.com
www.dimeye.com

DimEye is a California based high-tech company providing innovative solutions to high accuracy 3DMeasurement into inaccessible/hazardous environments.

Dive 1st Aid.................................... 633
599 W. 7th Street
San Pedro, CA 90731
310-832-4363
marketing@dive1staid.com
www.dive1staid.com

We manufacture some of the best first aid kits in the world. We are also a full medical supplier.

Exhibitor Directory

Divers Supply Inc.......................... 719
2396 Belle Chasse Highway
Gretna, LA 70056
504-392-2800 • F: 504-392-3920
tterrebonne@diverssupplyinc.com
www.diverssupplyinc.net

A commercial diving equipment supplier. Having been in business for over 29 years, located in the New Orleans and well suited for international sales.

Diving Unlimited International, Inc........................... 239
1148 Delevan Drive
San Diego, CA 92102
619-236-1203 • F: 619-237-0378
cmheaton@DUI-Online.com
www.dui-online.com

Draeger.. 745
7256 South Sam Houston
Parkway Suite 100
Houston, TX 77085
800-437-2437
www.draeger.com

Dräger is a technology leader in the fields of medical and safety technology, with over 105 years of experience in diving technology.

Drass Galeazzi................................ 744
Via P.N. Magri 112
Livorno 57121
Italy
0039 0586421221
federica.leonardo@drass.it
www.drass.it

Drass, the leader in the diving industry.

Dryden Diving Company, Inc. 347
213 Russell Mill Road
Woolwich Township, NJ 08085
856-467-1385 • F: 856-467-5771
drydiv@hotmail.com
www.DrydenDiving.com

We provide proprietary AWS D3.6:M 2017 compliant weld procedures, Diver Welders and Diver/CWI's to work with you.

E

ECA Group....................................... 344
262 Rue des Frères Lumière Z.I.
Toulon Est
La Garde 83130
France
33-494-089000 • F: 33-494-089070
ecagroup@ecagroup.com
www.ecagroup.com

The ECA Group is renowned for its expertise in robotics, aerospace & simulation to perform missions in hostile or restrictive environments.

EdgeTech... 740
4 Little Brook Road
West Wareham, MA 02576
508-291-0057 • F: 508-291-2491
amy.larose@edgetech.com
www.edgetech.com

EdgeTech is the leader in underwater technology solutions including acoustic releases, USBL systems, sidescan, sub-bottom profilers, bathymetry system.

Epic Divers & Marine **432**
10656 Highway 23
Belle Chasse, LA 70037
504-681-1205
cpa@epiccompanies.com
www.epiccompanies.com

EPIC Diving & Marine Services, LLC, a TETRA Company, provides commercial diving and marine services in the Gulf of Mexico and around the globe.

F

Fisk Marine Insurance International, LLC **446**
8110 Breakwater Drive
New Orleans, LA 70124
504-302-4400
ask@fiskusa.com
www.FiskUSA.com

ALL TYPES OF INSURANCE for Diving Contractors.

Five Star Products, Inc. **342**
60 Parrott Drive
Shelton, CT 06484
203-336-7900 • F: 203-336-7930
marketing@fivestarproducts.com
www.fivestarproducts.com

PileForm™ jacket systems and cementitious and epoxy fill material designed for the rehabilitation and stabilization of underwater structures.

Fugro **540**
6100 Hillcroft
Houston, TX 77081
713-369-5600
hhollis@fugro.com
www.fugro.com

Fugro proudly provides geo-intelligence and asset integrity solutions to contribute to a safe and liveable world.

G

Glenair Inc **734**
1211 Air Way
Glendale, CA 91201
818-247-6000
sales@glenair.com
www.glenair.com

Glenair manufactures high-performance interconnect solutions for harsh environmental applications. Our innovative products covers numerous markets.

Greensea **420**
10 E. Main Street
Richmond, VT 05477
802-434-6080
info@greensea.com
www.greensea.com

Navigation, Autonomy, UI, for ROV, AUV, Diver Vehicles and Surface Vehicles powered by patent-pending OPENSEA operating platform.

Gulf Engine & Equipment, Inc..... **621**
2306 Engineers Road
Belle Chasse, LA 70037
les@gulfengine.com
www.gulfengine.com

Exhibitor Directory

H

HeavyMetalDivers........................645
2189 Schwab Road
Marion, TX 78124
281-536-5294
info@heavymetaldivers.com
www.heavymetaldivers.com

Hydro Group Systems, Inc. 746
2188 Palm Way
Largo, FL 33771
727-812-0124
bmildon@hydrogroupsystems.com
www.hydrogroupsystems.com

Our focus is subsea cable/connector innovation.

Hydrospace Group Inc.................421
9559 Center Avenue, Suite P
Rancho Cucamonga, CA 91730
909-989-7773 • F: 909-912-8397
wkohnen@hydrospacegroup.com
www.hydrospacegroup.com

Design & Manufacturer of PVHO acrylic windows for Dive & Medical HBOT chambers. Design & Mfg of Hi-Rel subsea brushless DC motors from 1-100HP.

I

Innerspace Corp............................421
1138 E. Edna Place
Covina, CA 91724
omar@innerspacethrusters.com
www.innerspacethrusters.com

International Special Risks 333
50 Salem Street, 3rd Floor
Building B
Lynnfield, MA 01940
781-295-0270 • F: 781-246-7833
steve@isr-insurance.com
www.isr-insurance.com

Diving and Marine insurance specialist.

Interspiro, Inc.................................737
10225 82nd Avenue
Pleasant Prairie, WI 53158
203-500-5055 • F: 262-947-9902
bengt.kjellberg@interspiro.com
www.interspiro.com

Surface supplied dive system, with communication, dive computer, dark water vision, with all information stored and transferred live to the surface.

Inuktun Incommand Robotics, LLC..................................834
5041 Spencer Highway #601
Pasadena, TX 77505
844-468-5886 • F: 250-729-8077
sales@inuktun.com
www.incommandrobotics.com

Inuktun InCommand Robotics together with Seamor Marine provide remotely operated systems that bridge the gap between people and the physical world.

J

JFD Ltd 338
Enterprise Drive, Westhill
Aberdeen AB32 6TQ
United Kingdom
01224 740145
enquiries@jfdglobal.com
www.jfdglobal.com

JFD is the leading underwater capability provider, serving the commercial & defence markets with diving, submarine and hyperbaric solutions & services.

K

Kirby Morgan Dive Systems, Inc. 326
1430 Jason Way
Santa Maria, CA 93455
805-928-7772 • F: 805-928-0342
kbarragan@kirbymorgan.com
www.kirbymorgan.com

Kirby Morgan, a name that has become synonymous with quality and cutting-edge design throughout the world.

M

Mactech Offshore 646
302 Ida Road
Broussard, LA 70518
337-839-2793 • F: 337-839-2794
arobichaux@mactechoffshore.com
www.mactechoffshore.com

From Split Frame Cutters to Diamond Wire Saws, we provide specialized cutting tools to the offshore oil industry.

Marine Sonic Technology/ Atlas North America 345
120 Newsome Drive, Suite H
Yorktown, VA 23692
804-693-9602 • F: 804-693-6785
regan.lipinski@na-atlas.com
www.marinesonic.com

Marine Sonic Technology a world leader in high resolution side scan sonar systems.

Marine Technology Reporter..... 541
118 East 25th Street
New York, NY 10010
212-477-6700 • F: 212-254-6271
howard@marinelink.com
www.marinetechnologynews.com

The industry's largest circulating magazine.

N

Northern Diver International, Ltd 647
Appley Lane North Appley Bridge
Lancs WNG 9AE
United Kingdom
0044-1257-256949
neil@ndiver.com
www.ndiver-commercial.com

Founded over 40 years ago, Northern Diver is proud to supply numerous commercial outfits around the world with specialist diving suits and equipment.

Nuvair 332
1600 Beacon Place
Oxnard, CA 93033
805-815-4044 • F: 805-486-0900
glenn@nuivair.com
www.nuvair.com

Nuvair provides innovative solutions for producing nitrox along with extensive offerings of low and high pressure breathing air compressors.

Exhibitor Directory

O

Ocean Technology Systems.......322
3133 W. Harvard Street
Santa Ana, CA 92704
714-754-7848
rlummus@otscomm.com
www.otscomm.com

OTS is the global leader in Underwater Communications & FFMs. Be sure to check out OTS' newest innovation: The MK-III Comms Box with Bluetooth.

O'Three Ltd......................................546
Osprey Quay
Portland DT5 1BL
United Kingdom
0044 1305-822820
F: 0044 1305-822821
sales@othree.co.uk
www.othree.co.uk

Manufacturers of quality dry suits, hot water suits and neoprene accessories for commercial, public safety and military and for over 30 years.

Outland Technology.....................526
38190 Commercial Court
Slidell, LA 70458
985-847-1104
jeff@outlandtech.com
www.outlandtech.com

Outland Technology is a leading designer and manufacturer of cutting-edge Underwater ROVs and other monitoring solutions for offshore applications.

P

Praxair Offshore Services..........735
1585 Sawdust Road, Suite 300
The Woodlands, TX 77380
800-PRAXAIR
info@praxair.com
www.praxair.com/services

Q

Quikrete Companies....................727
Five Concourse Parkway, Suite 1900
Atlanta, GA 30328
985-227-5373 • F: 601-798-6022
tgrazier@quikrete.com
www.slingbag.net

Slingbag is the new STANDARD in UNDERWATER PIPELINE SUPPORT providing a SAFE and effective means to deliver 56-60 lb burlap bags to the project site.

R

ROV Planet, Ltd..............................544
49 Esslemont Avenue
Aberdeen AB25 1ST
United Kingdom
+447868023876
info@rovplanet.com
www.rovplanet.com

The most ROV focused publication in the world for the ROV/AUV industry. 2,000 printed and 12,500 pdf copies are sent to the relevant industry players.

Exhibitor Directory

Rovsco, Inc. **620**
5263 Barker Cypress Road, Suite 600
Houston, TX 77084
281-858-6333 • F: 281-858-6363
sales@rovsco.com
www.rovsco.com

Established in 1986, Rovsco is one of the leading companies specialized in supply & support, focused toward Work-Class ROV operators and Com. Diving.

S

Saab **641**
2101 L Street, NW Suite 350
Washington, DC 20037
703-406-7209
paul.bryson@saabgroup.com
www.saab.com

Saab Seaeye's vehicles range from the portable and versatile Seaeye Falcon to the revolutionary electric work class Seaeye Jaguar.

Sea Technology Magazine **642**
1600 Wilson Boulevard, Suite 1010
Arlington, VA 22209
703-524-3136 • F: 703-841-0852
seatechads@sea-technology.com
www.sea-technology.com

Published monthly for 54+ years. Sea Technology magazine is available in print and digitally and is circulated in more than 110 countries.

SEACON **444**
1700 Gillespie Way
El Cajon, CA 92020
619-562-7071 • F: 619-562-9706
melanie.harrison@te.com
www.seaconworldwide.com

TE Connectivity (TE)'s SEACON portfolio includes over 2500 underwater electrical and fiber optic connectors, and complete connectivity systems.

Exhibitor Directory

SeaDrone **233**
1045 Noel Drive, Suite 3
Menlo Park, CA 94025
928-257-2826
www.seadronepro.com

Seanamic Group - Umbilicals International **440**
10711 Cash Road
Stafford, TX 77477
281-275-6600
info@umbilicals.com
www.umbilicals.com

Simpson Strong-Tie **733**
5956 Las Positas Boulevard
Pleasanton, CA 94588
800-925-5099
femerson@strongtie.com
www.strongtie.com

Simpson Strong-Tie® Repair, Protection and Strengthening Systems products are designed, tested, and manufactured to provide lasting results.

Southwest Research Institute... **337**
6220 Culebra Road
San Antonio, TX 78238
210-522-5410
alan.steiner@swri.org
www.swri.org

Benefiting government, industry and the public through innovative science and technology.

Subsalve USA Corporation **520**
P.O. Box 2030
North Kingstown, RI 02852
401-884-8801 • F: 401-884-8868
richard@subsalve.com
www.subsalve.com

Subsalve USA is the world leading innovator in buoyancy & engineered inflatables delivering the best products available all manufactured in the USA!

Subsea Technologies, Inc. **627**
1323 Price Plaza Drive
Katy, TX 77449
281-398-5600
sales@subseatechnologies.com
www.subseatechnologies.com

Sales, rentals, and service of GPS positioning systems, ROV instrumentation, side scan and sub-bottom systems, and rugged field computing solutions.

SULIS Subsea Corporation **547**
4-17 Dundee Avenue
Mount Pearl, NL A1N 4R6
Canada
709-745-5080
info@sulissubsea.com
www.sulissubsea.com

Canadian developer of 4K UHD camera systems with spectacular image quality for underwater applications.

Survey Equipment Services **327**
1775 Westborough Drive
Katy, TX 77449
281-829-6481
acraig@ses-services.com
www.ses-services.com

Come visit us and our Co-Exhibitors on Booth 327 Sonardyne Inc, DOTgroup International Blue Ocean Monitoring.

T

Trelleborg Offshore Boston........747
58 Teed Drive
Randolph, MA 02368
774-719-1433
brianna.orourke@trelleborg.com
www.trelleborg.com

Trelleborg Offshore Boston, part of Trelleborg Applied Technologies, manufactures and develops engineering polymer and syntactic material.

U

UBC Pile Drivers and Divers......543
750 Dorchester Avenue
Boston, MA 02125
617-443-1988

UBC Divers and Pile Drivers represents over 1000 skilled commercial divers and numerous marine construction contractors throughout the US and Canada.

W

Water Linked.................................233
Stiklestadveien 1
Trondheim 7041
Norway
+47-406-03151
post@waterlinked.com
www.waterlinked.com

Water Linked is disrupting the market for underwater acoustics with cost effective positioning systems for ROV's and divers.

WT Industries.................................644
11328 Neeshaw Drive - Unit C
Houston, TX 77065
832-731-7098
dtrail@wt-ind.com
www.wt-ind.com

WT Industries, LLC is an innovative engineering technology company specializing in launch and recovery systems for offshore equipment.

Z

Zhejiang Jianzhong Maritime Engineering Equipment Co., Ltd........................743
Floor 1, Building 8 Wuzhen
Industrial Zone, Wuzhen Town
Tongxiang
Jiaxing Zhejiang Province 314501
China
0086-573-88715777
lumin@oilabsorbspill.com
www.oilabsorbspill.com

JXY is specialized in manufacturing, designing and technical service support of Silt Curtain and Oil Spill Response Equipments from China.

Exhibitor Product Categories

Acoustic Positioning
Aqua Tech Solutions.................... 321
Northern Diver International, Ltd........................ 647
Subsea Technologies, Inc........... 627
Survey Equipment Services....... 327
Water Linked................................. 233

Actuators - Linear & Rotary
Hydrospace Group Inc................. 421

Anchors
Boats & Harbors........................... 742
Simpson Strong-Tie...................... 733

Autonomous Underwater Vehicles
BIRNS, Inc...................................... 720
ECA Group...................................... 344
Saab.. 641
Trelleborg Offshore Boston........ 747

Bridges & Dams
Boats & Harbors........................... 742
UBC Pile Driver and Divers (Pile Drivers Local 56)............... 543
WT Industries................................ 644

Cables, Hoses & Umbilicals
AK Industries................................. 427
Anekonnect Inc............................ 545
Aqua Tech Solutions.................... 321
Armada Systems, Inc.................. 441
AXSUB Inc..................................... 737
Bay-Tech Industries, Inc............ 443
Drass Galeazzi.............................. 744
Hydro Group Systems, Inc......... 746
Interspiro, Inc................................ 737
Nuvair.. 332
Rovsco, Inc.................................... 620
SEACON... 444

Cameras/ Photography/Monitors
AXSUB Inc..................................... 737
Darkwater Vision.......................... 737
DeepSea Power & Light, Inc...... 542
Dimensional Eye........................... 335
Interspiro, Inc................................ 737
Rovsco, Inc.................................... 620
SULIS Subsea Corporation......... 547

Cavidyne Blasters
Aqua Tech Solutions.................... 321
Rovsco, Inc.................................... 620

Commercial Diving
BIRNS, Inc...................................... 720
Boats & Harbors........................... 742
Cochran Consulting, Inc............. 737
Cylinder Training Services.......... 732
EPIC Divers & Marine.................. 432
Kirby Morgan Dive Systems, Inc............................... 326
UBC Pile Driver and Divers (Pile Drivers Local 56)............... 543

Commercial Diving Consultants
UBC Pile Driver and Divers (Pile Drivers Local 56)............... 543

Commercial Diving Contractors
Boats & Harbors........................... 742
Quikrete Companies..................... 727
UBC Pile Driver and Divers (Pile Drivers Local 56)............... 543

Commercial Diving Equipment Rentals
Bay-Tech Industries, Inc............ 443
Northern Diver International, Ltd........................ 647

Exhibitor Product Categories

Commercial Diving Equipment Sales

Aqua Lung America....................626
Aqua-Air Industries, Inc. 521
Armada Systems, Inc...................441
Avon Protection Systems............436
AXSUB Inc 737
Bay-Tech Industries, Inc. 443
Boats & Harbors..........................742
Cochran Consulting, Inc. 737
Cylinder Training Services 732
David Clark Company, Inc. 637
Dive 1st Aid 633
Divers Supply Inc. 719
Drass Galeazzi...............................744
Draeger..745
Hydrospace Group Inc.................421
Interspiro, Inc. 737
JFD Ltd 338
Kirby Morgan Dive Systems, Inc. 326
Northern Diver International, Ltd 647
Nuvair...332
Rovsco, Inc. 620
Subsalve USA Corporation.........520
Water Linked 233

Commercial Diving Schools

Cylinder Training Services..........732
Northern Diver International, Ltd 647

Compressors

Aqua Tech Solutions.....................321
Aqua-Air Industries, Inc. 521
Armada Systems, Inc 441
Divers Supply Inc. 719
Draeger...745
Northern Diver International, Ltd 647
Nuvair...332

Control Systems

Greensea..420
Hydrospace Group Inc.................421
JFD Ltd 338
Northern Diver International, Ltd 647
Subsea Technologies, Inc...........627

Corrosion Control & Sealing

Armada Systems, Inc 441
Denso North America 527
Rovsco, Inc. 620
Simpson Strong-Tie 733
Trelleborg Offshore Boston 747

Exhibitor Product Categories

Decommissioning
Aqua Tech Solutions.......321
Fugro.......540
Mactech.......646

Diver Communications
Aqua-Air Industries, Inc.......521
AXSUB Inc.......737
Cochran Consulting, Inc.......737
David Clark Company, Inc.......637
Greensea.......420
Interspiro, Inc.......737
JFD Ltd.......338
Water Linked.......233

Diving Helmets
Aqua Lung America.......626
Bay-Tech Industries, Inc.......443
Divers Supply Inc.......719
JFD Ltd.......338
Kirby Morgan Dive Systems, Inc.......326

Diving Suits
Aqua Lung America.......626
Avon Protection Systems.......436
Bay-Tech Industries, Inc.......443
Divers Supply Inc.......719
JFD Ltd.......338
Northern Diver International, Ltd.......647
O'Three Ltd.......546

Engineering
Anekonnect Inc.......545
Avon Protection Systems.......436
Beta Subsea Systems.......318
Cochran Consulting, Inc.......737
Deep Ocean Engineering, Inc.......341
Fugro.......540
JFD Ltd.......338
Rovsco, Inc.......620
Southwest Research Institute.......337

Equipment Repair & Refurbishment
Cylinder Training Services.......732
JFD Ltd.......338
Subsea Technologies, Inc.......627
WT Industries.......644

Excavation Equipment
Aqua Tech Solutions.......321

Fiberglass Forms
Denso North America.......527

Fibre Optic Cables
BIRNS, Inc.......720
Glenair Inc.......734
Hydro Group Systems, Inc.......746
SEACON.......444

Grouts & Adhesives
Denso North America.......527
Five Star Products, Inc.......342
Simpson Strong-Tie.......733

Gyro Compass
Aqua Tech Solutions.......321
Subsea Technologies, Inc.......627
Survey Equipment Services.......327

Health & Safety Equipment
Cochran Consulting, Inc.......737
Divers Supply Inc.......719
Nuvair.......332
Rovsco, Inc.......620
Zhejiang Jianzhong Maritime Engineering Equipment Co., Ltd.......743

Hot Water Heaters
Aqua-Air Industries, Inc.......521
Bay-Tech Industries, Inc.......443
Divers Supply Inc.......719

Exhibitor Product Categories

Hydraulic Pressure Units
Aqua Tech Solutions.................... 321
Armada Systems, Inc.................. 441
Bay-Tech Industries, Inc. 443
Boats & Harbors........................... 742
Divers Supply Inc. 719
Rovsco, Inc. 620

Industrial & Specialty Gas
Nuvair.. 332

Inertial Navigation Systems
Greensea 420
Survey Equipment Services 327

Infection Control
Dive 1st Aid 633

Infection Prevention
Dive 1st Aid 633

Insurance
Fisk Marine Insurance International, LLC 446

Lift Bags
Aqua Tech Solutions.................... 321
Boats & Harbors........................... 742
Divers Supply Inc. 719
Northern Diver International, Ltd 647
Quikrete Companies 727
Subsalve USA Corporation 520

Manipulators
ECA Group....................................... 344

Marine Construction
Boats & Harbors........................... 742
David Clark Company, Inc. 637
EPIC Divers & Marine 432
Mactech 646
Quikrete Companies 727
Southwest Research Institute ... 337
UBC Pile Driver and Divers (Pile Drivers Local 56) 543
Zhejiang Jianzhong Maritime Engineering Equipment Co., Ltd. 743

Media & Publications
Boats & Harbors........................... 742
Marine Technology Reporter..... 541
ROV Planet, Ltd. 544
Sea Technology Magazine 642

Exhibitor Product Categories

Motion Sensing Equipment
Survey Equipment Services 327

Offshore Oil & Gas
Anekonnect Inc. 545
Avon Protection Systems 436
BIRNS, Inc. 720
Broco Inc. 723
David Clark Company, Inc. 637
Dimensional Eye 335
ECA Group 344
Glenair Inc 734
JFD Ltd 338
Nuvair 332
Saab 641
Southwest Research Institute ... 337
Subsea Technologies, Inc. 627
Trelleborg Offshore Boston 747
WT Industries 644
Zhejiang Jianzhong Maritime Engineering Equipment Co., Ltd. 743

Petrolatum Wrap Systems
Denso North America 527

Pile Repair/Pile Cleaning
EPIC Divers & Marine 432
Five Star Products, Inc. 342
UBC Pile Driver and Divers (Pile Drivers Local 56) 543

Propulsion Systems
Boats & Harbors 742
Hydrospace Group Inc 421

Regulations and Safety
Cylinder Training Services 732
Dive 1st Aid 633

Remotely Operated Vehicles
BIRNS, Inc. 720
Blue Robotics, Inc. 640
David Clark Company, Inc. 637
Deep Ocean Engineering, Inc. 341
Digital Edge Subsea 644
ECA Group 344
Greensea 420
Outland Technology 526
Rovsco, Inc. 620
Saab 641
Survey Equipment Services 327
Trelleborg Offshore Boston 747
WT Industries 644

ROV Operators/Contractors
David Clark Company, Inc. 637
Deep Ocean Engineering, Inc. 341
ECA Group 344
UBC Pile Driver and Divers (Pile Drivers Local 56) 543
WT Industries 644

Saturation Diving Systems
Aqua-Air Industries, Inc. 521
Bay-Tech Industries, Inc. 443
David Clark Company, Inc. 637
Divers Supply Inc. 719
Drass Galeazzi 744
Hydrospace Group Inc 421
JFD Ltd 338

Security & Surveillance Systems
David Clark Company, Inc. 637

Sonar & AUV Surveys
Dimensional Eye 335
ECA Group 344
EdgeTech 740

Sub Sea Cutting Tools
Aqua Tech Solutions 321
Aqua-Air Industries, Inc. 521
Northern Diver International, Ltd 647
Trelleborg Offshore Boston 747

Exhibitor Product Categories

Thrusters - ROV, AUV
Hydrospace Group Inc. ... 421

Training & New Technologies
Avon Protection Systems ... 436
Drass Galeazzi ... 744
ECA Group ... 344
JFD Ltd ... 338
Marine Sonic Technology/ Atlas North America ... 345
Saab ... 641

Ultrasonic Thickness Gauges
Aqua-Air Industries, Inc. ... 521
Cygnus Instruments, Inc. ... 736

Underwater Flotation
Subsalve USA Corporation ... 520
Trelleborg Offshore Boston ... 747

Underwater Imaging Systems
Anekonnect Inc. ... 545
AXSUB Inc ... 737
BIRNS, Inc. ... 720
Darkwater Vision ... 737
Digital Edge Subsea ... 644
Dimensional Eye ... 335
Interspiro, Inc. ... 737
Marine Sonic Technology/ Atlas North America ... 345
Northern Diver International, Ltd ... 647
Outland Technology ... 526
Subsea Technologies, Inc. ... 627
SULIS Subsea Corporation ... 547
Survey Equipment Services ... 327

Underwater Inspection
Aqua-Air Industries, Inc. ... 521
AXSUB Inc ... 737
BIRNS, Inc. ... 720
Darkwater Vision ... 737
Deep Ocean Engineering, Inc. ... 341
Dimensional Eye ... 335
Dryden Diving Company, Inc. ... 347
ECA Group ... 344
Fugro ... 540
Greensea ... 420
Saab ... 641
Subsea Technologies, Inc. ... 627
WT Industries ... 644

Underwater Inspection Equipment
Anekonnect Inc. ... 545
Aqua-Air Industries, Inc. ... 521
Bay-Tech Industries, Inc. ... 443
BIRNS, Inc. ... 720
Blue Robotics, Inc. ... 640
Cochran Consulting, Inc. ... 737
Cygnus Instruments, Inc. ... 736
Darkwater Vision ... 737
Digital Edge Subsea ... 644
Dimensional Eye ... 335
Interspiro, Inc. ... 737
Nuvair ... 332
Outland Technology ... 526
Saab ... 641
Subsea Technologies, Inc. ... 627
Survey Equipment Services ... 327

Underwater Lighting
Anekonnect Inc. ... 545
AXSUB Inc ... 737
BIRNS, Inc. ... 720
Blue Robotics, Inc. ... 640
Darkwater Vision ... 737
DeepSea Power & Light, Inc. ... 542
Drass Galeazzi ... 744
Outland Technology ... 526
Subsea Technologies, Inc. ... 627
Survey Equipment Services ... 327

Exhibitor Product Categories

Underwater Structural Welding Services
Dryden Diving Company, Inc. 347

Unmanned Surface Vessels
Deep Ocean Engineering, Inc. 341
ECA Group 344
Greensea 420
Survey Equipment Services 327

Welding & Burning Equipment
Aqua-Air Industries, Inc. 521
Bay-Tech Industries, Inc. 443
Broco Inc. 723
Divers Supply Inc. 719

Welding Equipment
Broco Inc. 723
Rovsco, Inc. 620

Welding Services
Dryden Diving Company, Inc. 347

Advertisers

ADVERTISER	WEBSITE	PAGE
Aqua Tech Solutions	www.aqua-techsolutions.com	32
Armada Systems, Inc	www.armadahull.com	43
Dryden Diving Company, Inc.	www.drydendiving.com	39
Kirby Morgan Dive Systems, Inc.	www.kirbymorgan.com	Outside Back Cover
Minnesota Commercial Diver Training Center	www.mncdtc.com	45
Novatec Braids Ltd.	www.novabraid.com	6
The Ocean Corporation	www.oceancorp.com	4

UNDERWATER INTERVENTION
2019
New Orleans, LA
February 5-7, 2019
SEE YOU NEXT YEAR!
UI2019
UNDERWATER INTERVENTION

UI2018
UNDERWATER
INTERVENTION

Cost savings using Manned Underwater Intervention in suitable water depths. Experience from the Norwegian Continental Shelf

Rolf Røssland
Managing Director
NUI AS

nui
By depth of experience

Agenda

Short introduction
Diving methods
Manned Underwater Operations offshore
Pro and Cons using divers

nui
By depth of experience

My Diving background

Diver – Supervisor – Diving Advisor – Manging Director

nui

UNIQUE COMPETENCE

Since 1976, NUI have delivered subsea services and competence to the oil and gas industry. With our expertise and unique facility, we are a solid partner for both national and international subsea operations.

nui
By depth of experience

NUI OWNER STRUCTURE

Members
- Total E&P Norge
- AkerBP (Det norske/BP)
- Repsol Norge (Talisman)
- Spirit Energy (Centrica)
- Statoil

Affiliated Members
- Point Recourse (ExxonMobil)
- Gassco

Associated Members
- Technip Norge
- Subsea 7 Norge

THE ASSOCIATION OF OPERATORS FOR HYPERBARIC LIFEBOAT RECEPTION FACILITIES

OFHB (Operatørselskapenes Forening for Hyperbar Livbåtberedskap)

nui

NUI main business area

- Hyperbaric Reception Centre for the North Sea region
- Testing of equipment, pressure integrity, chemical composition, hyperbaric capacity and physical fitness.
- Advisory services for the offshore industry
- Training of personnel
- HBO Treatment

www.nui.no

nui

Manned Underwater Operation (Diving)

nui
By depth of experience

MUO

- For working underwater there are two main intervention methods to be used, MUO (Diving) and ROV.
- Both methods has their pro and cons that needs to be evaluated when deciding what intervention method to use.
- It is important to remember that today the ROV is an important part of most diving operations and it's used for both support, operational work and for safety purposes.

nui

325 BC Depth 25 msw

1535

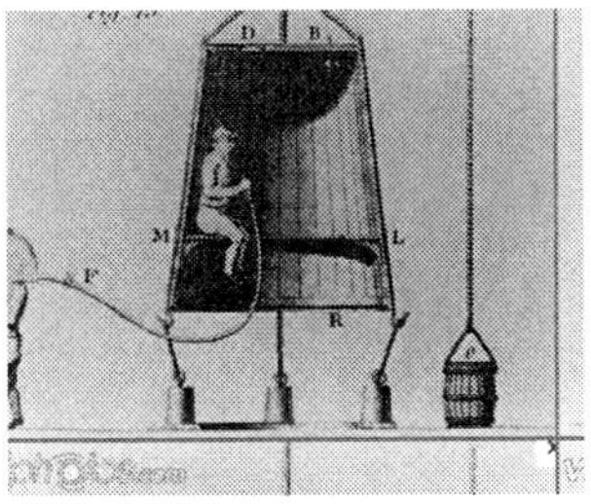

Edmond Halley's diving bell of 1716

1797

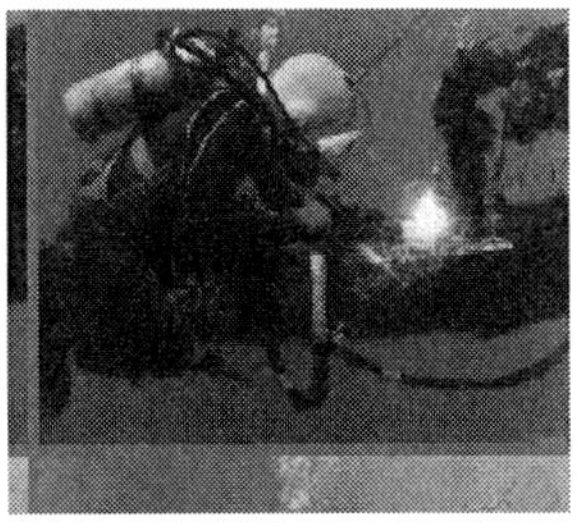

MUO

- Only reason to dive offshore is to perform a work task
- Only reason to use divers to solve the work task is that it's the most cost effective way to do the work.
- If there are proper safety focus, diving is a very safe and efficient method to intervene under water.
- Normally the deep diving (sat diving) is in combination with ROV

MUO

There are several methods and types of diving. The four types of methods relevant for diving at the Norwegian Continental Shelf (NCS) **was**:

- Self contained diving (SCUBA)
- Surface supplied diving
- Saturation diving.
- Atmospheric Diving Suits (ADS)

Today on the *NCS* we only use

- Surface supplied diving and Saturation diving
  and the Saturation diving normally includes use of ROVs

MUO - Safety

Diving is considered to be a safe intervention method by the Norwegian Petroleum Safety Authority (PSA). They state:

- *Diving on the NCS is conducted on acceptable terms and at a level of risk comparable with other petroleum industry in Norway. These conditions have prevailed for longer than many might think.*
  - *no deaths related to diving in the petroleum industry since 1987*
  - *few serious personal injuries or cases of the bends over the past two decades, but some serious near-misses – particularly in subsea lifting operations*
  - *substantial improvements in equipment and the working environment over the past 30 years.*

MUO - Safety

Diving is also considered to be a safe and efficient intervention method from most Operators in the North Sea Area.

Examples: Statoil, Total, ConocoPhillips, AkerBP, Repsol, BG Norge, BP, Shell and others are conducting MUO's on a regular basis.

Statoil strategy related to Diving is that this intervention method shall be considered for all underwater operations on relevant water depths.

The most cost-effective and operational suitable method shall be used.

They state:

- Manned underwater operations can be the preferred solution based on evaluation of safety, efficiency and cost. *And not only a solution to be chosen where remote operations cannot perform the specified task*

Diving – Surface Supplied

- Surface supplied Diving is a flexible intervention method suitable for work between surface and approx. 20 msw.
- There are special designed small diving vessels fit for the purpose, but other solutions are also used.
- It is normally operated from a suitable support vessel with the handling system for the diving boat and a chamber system onboard.
- The intervention method is regularly used offshore.

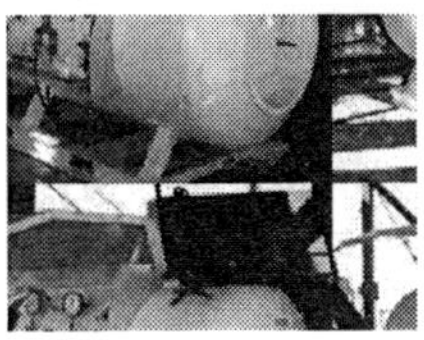

Diving – Saturation Diving

Saturation diving allows the divers to live and work at pressures for days or weeks at a time (regulatory limitations). This type of diving allows for greater economy of work and enhanced safety for the divers.

After the bellrun the divers rest and live in a pressurized chamber system onboard the diving support vessel (DSV), or other floating work station, at approximately the same pressure as the work depth.

The diving team is compressed to the working pressure only once, at the beginning of the work period, and decompressed to surface pressure once, after the entire work period of days or weeks.

Diving – Saturation Diving

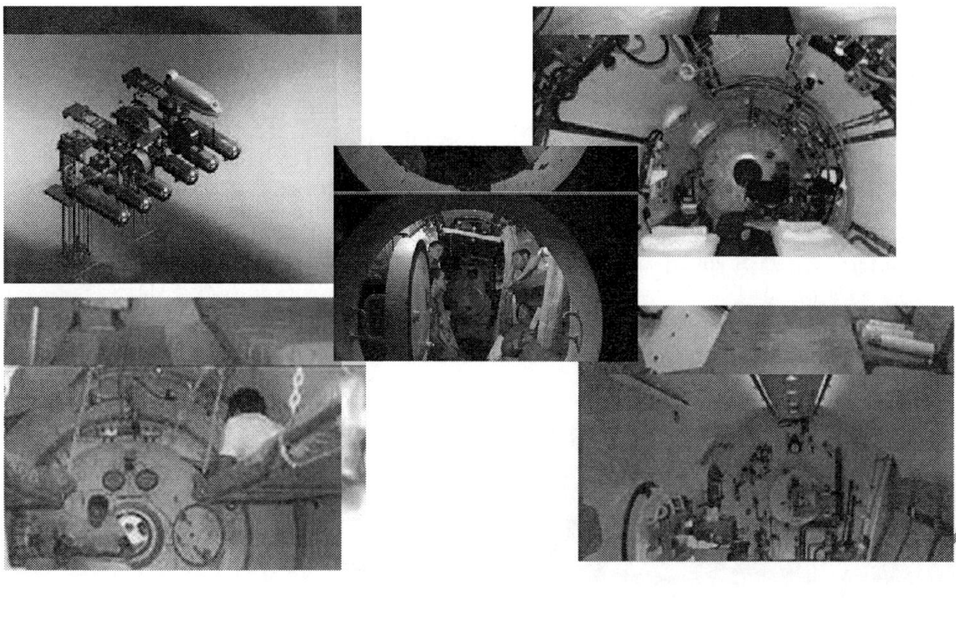

Diving – Saturation Diving

Bell is lifted from the chamber system and lowered through the moon pool

nui

Diving – Saturation Diving

Diver is locked out from the bell and work in water or habitat.

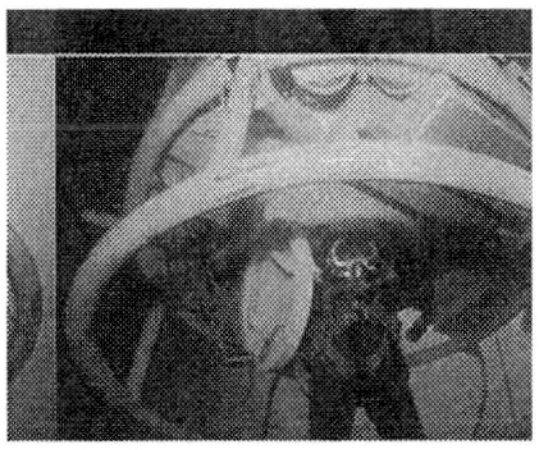

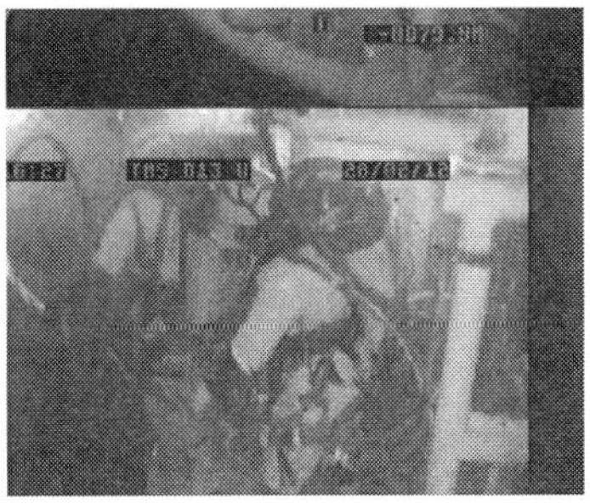

nui

Diving

What can a diver do underwater?

The diver can perform (almost) all the work tasks that a man can do on the surface – its only takes a little more time and planning.

- Inspection
- Construction (tie-in, connectors, cutting, mounting, lifting, etc)
- Welding (wet and dry)
- Maintenance
- Grinding
- Cleaning

Diving – Pro

- **Tried and tested** method in water depths up to 250m (Special attention passing 180)
- Normally **no specialist tooling required**
  - Special tooling leads to more design, more planning, higher cost.
  - Hubs are generally longer lead time than flanges
  - Operations can be conducted with standard off-the-shelf equipment
- **Cost effective**
  - More robust regarding schedule in case of deviations in engineering
  - Less costly project specific equipment required
  - Flexible and problem solving method
- **Keeps size/weight down**
  - On umbilical terminations
  - On flow/guide bases
- **Accessibility**
  - Access to places where ROV and tooling is to big
- **Problems** are easier to spot / **resolve**
  - Dexterity / vision of divers is far greater than ROV
  - Rigging problems can be solved using low complexity equipment

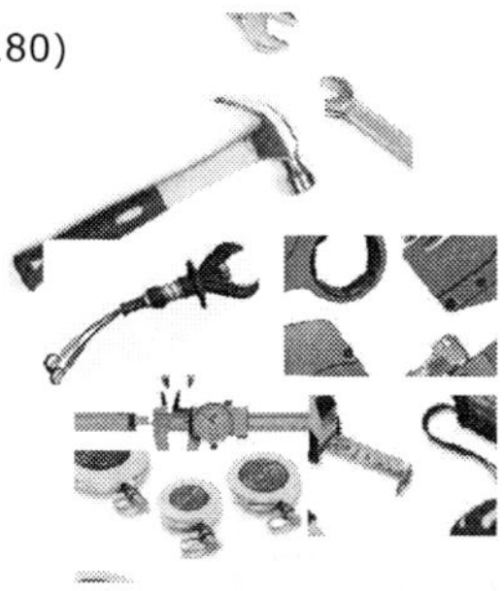

Diving – Cons

- High focus from authorities (PSA in Norway)
  - Some extra work <u>but also a PRO</u>, increase the focus on safety both during planning and execution)
- Restriction on depth 0 – approx. 250 msw (and deeper)
- Operational restrictions
  - Time in water (regulatory)
  - Time in saturation (regulatory)
  - Working distance from bell and diving station
  - Upwards and downwards excursions
- Stand alone operations might entail extra vessel days caused by decompression.
  - Example :Living on 90 msw, Decomp approx 4,5 days. (plus stabilisation time)

nui

How to capitalize on Diving?

- Some of the value of diving lies in the simplification that can be achieve in the seabed facilities.
  - **Simpler design – less steel (no pull-in forces)**
  - **Reduced dimensions on spools**
  - **Less protection covers**
  - **No advanced connectors and tie-in tooling**
  - **No PLET/PLEMs required**
  - **Reduced lifting weights / simplified installation**
  - **Easier access for divers in structures**
- Invite marine contractors to integrate diving in bids
- Modern DSV's are state-of-the-art construction vessels, not only for diving.

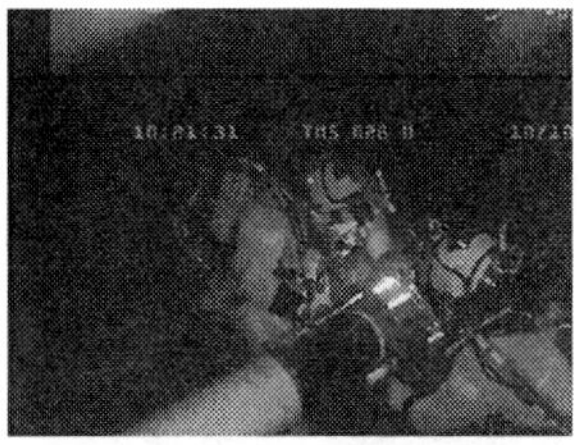

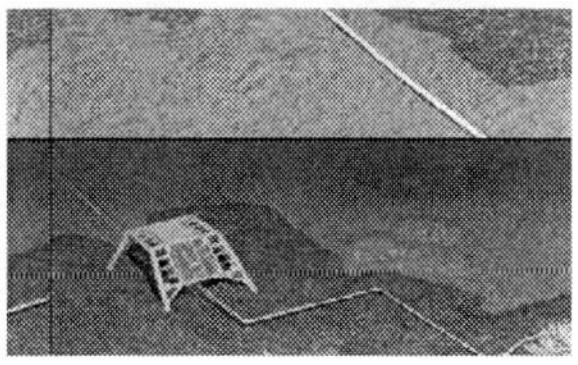

nui

How to capitalize on Diving? Example

- Diver tie backs = Low Capex cost
- Diver flanges = Smaller spools = Can be installed by diving vessels

Diver connections vs Remote connections

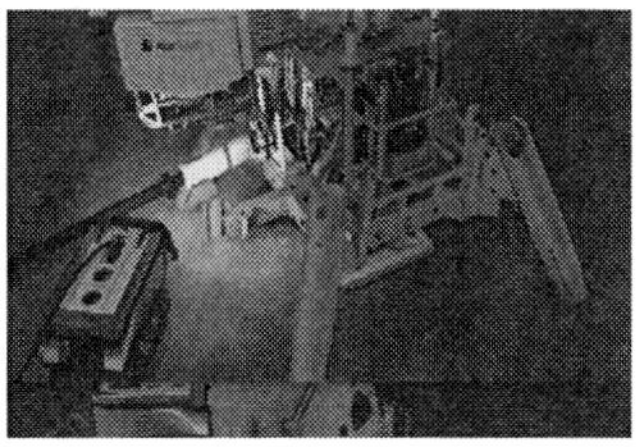

Diving Vessel Capabilities – more than just diving

Modern Diving Support Vessels (DSV) are in fact well equipped subsea construction vessels, with an additional tool – the divers.

DEEP ARCTIC (Skandi Arctic)

- Deck area 1700m^2
- 400T SWL subsea crane
- 2 WROVs + OBSROV
- 24 men diving system
- 2 diving bells

SEVEN FALCON

- Deck area 1050m^2
- 250T SWL subsea crane
- 1 WROV + OBSROV
- 24 men diving system
- 2 diving bells (4 men bellrun capacity)

Reduced Cost and Added Value by Diving

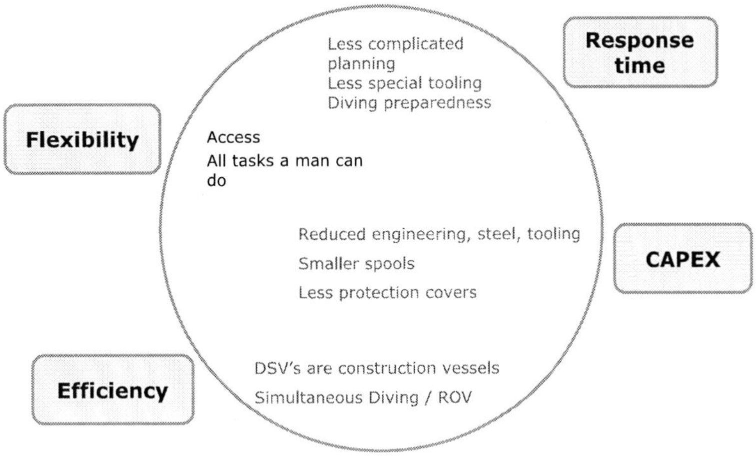

Why Diving - Example

There are several examples that use of MUO is a cost-effective intervention method.

- In the Statoil project "M10-80" there was identified production errors on 10mm bolts that attached pressure sensors on sub-sea installations. The project includes installations both on and off MUO limits (depth) and gave very good indications of the cost and efficient for both divers and ROVs.

Why Diving – Example "M10-80" Statoil project

- *The development of ROV tools had a higher cost that the whole MUO operation.*
- *The total MUO was 35 days operation / covered 50% of the total work.*
- *Project cost (*1000MNOK) *:*
  - ***The cost for the MUO was approx. 10% of the total project cost and covered more than 50% of the work.***
- *The divers changed the bolts and **repaired the failure**.*
- *ROV installed clamps.*
- *Divers benefit: management of change, flexible solutions, access, standard tools, etc.*

nui

Example of diving efficiency – Removal operations, Statoil project

SOW	Interventions method	Time	Cost
Statfjord SPM* removal	WROV	56 days	
GF SPM 1 removal (2014)	Diving + WROV	14 days	50% reduction
GF SPM 2 removal (2014)	Diving + WROV	20 days	30% reduction

- Statfjord SPM removal was a cut operation with no re-use of subsea installations.
- Gullfaks SPM removals were more complex with re-use of subsea assets (OLS installation)

*SPM = Single Point Mooring (oil loading buoy)

nui

Example (Study) of diving CAPEX reduction – Johan Sverdrup

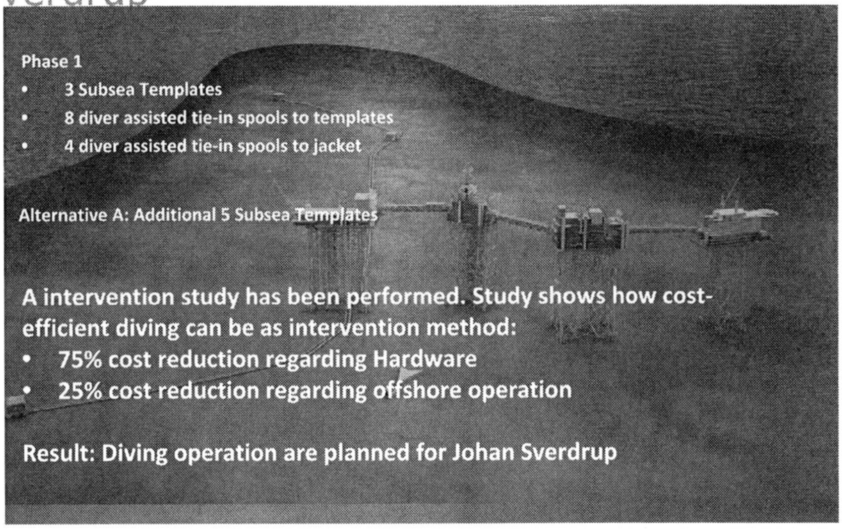

nui

Other Examples

- TOTAL E&P Norge – Martin Linge
- BG Norge – Gaupe
- Other Operators at NCS are planning for MUO
- Operators worldwide are performing MUO on a regular basis.

nui

Last initiative for more efficient operations - use of 4 -men bell run

- February 2016 Statoil used DSV Seven Falcon for Bundle repair at Gullfaks towhead C3
  - DSV Seven Falcon is the only DSV vessel designed and built for 4 men bell run capacity.
  - For this operation it was the first time a 4-men bell run was carried out and thereby utilized the fully capacity at DSV Seven Falcon
- This bell-runs showed a clear increasing of the efficiency for the Gullfaks projects
- When using one extra diver in the water, the generic risk increased. The risk was compensated with familization and active use of ROV if an incident should occurre.

nui

Conclusion

- If we accept to perform work using people – diving is as safe as other work-tasks if properly performed.
- The benefits of using manned underwater operations must be considered in the early phase of subsea projects.
  - Invite marine contractors to integrate diving in bids – if not, MUO will not be presented as an option
- In all phases in project planning;
  - Do not define concept or contract strategy in such a way that it's exclude use of manned intervention.
  - Keep the option open for manned operations in the ITT phase to harvest the value.
- This work can only be done due to good HSE statistics and based on thorough risk management

nui

Unmanned is not always the best solution

Wednesday, 06 April, 2016, 12:42pm
ednesday, 06 April, 2016, 2:20pm

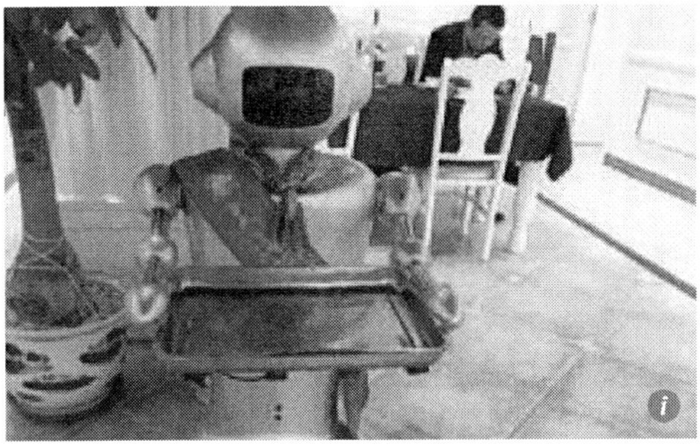

Employing artificial intelligent robots in Chinese restaurants – an idea that attracted national headlines – has not proved such a smart idea after all, mainland media reports.

From South China Morning Post

nui

Unmanned is not always the best solution

From Bloomberg Business

Mercedes Boots Robots From the Production Line

- "Robots can't deal with the degree of individualization and the many variants that we have today,"
- "The variety is too much to take on for the machines,"
- "They can't work with all the different options and keep pace with changes."

http://www.bloomberg.com/news/articles/2016-02-25/why-mercedes-is-halting-robots-reign-on-the-production-line

nui

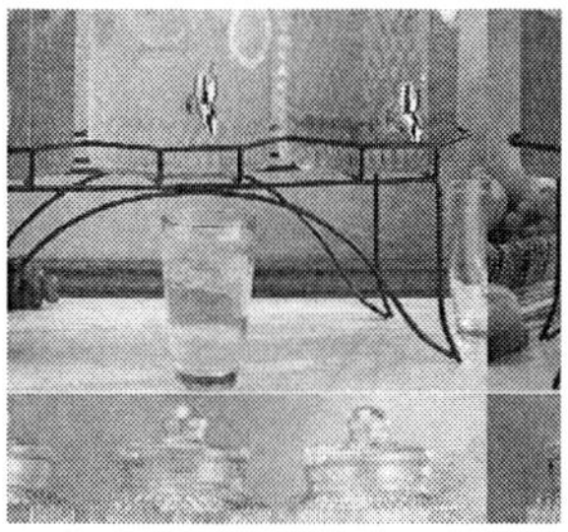

High tech solutions is often possible – but not always cost-effective or smart.

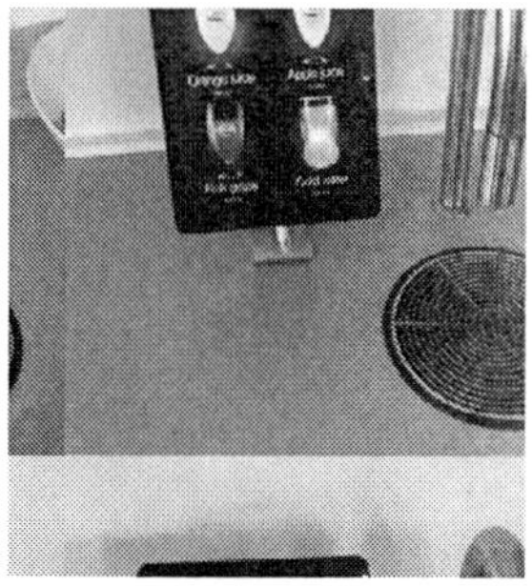

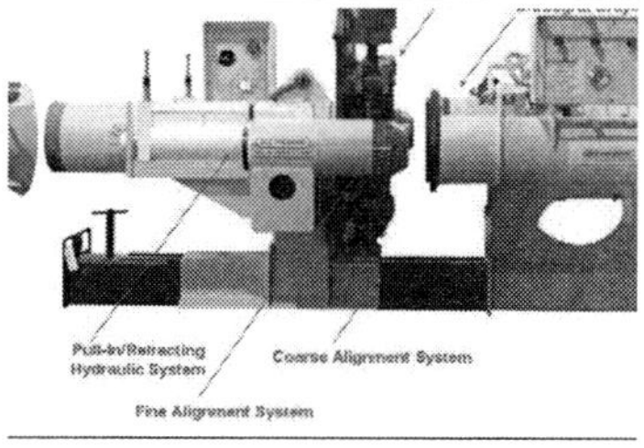

Thank you for your attention.

UI2018 UNDERWATER INTERVENTION

CITF OVERVIEW
1

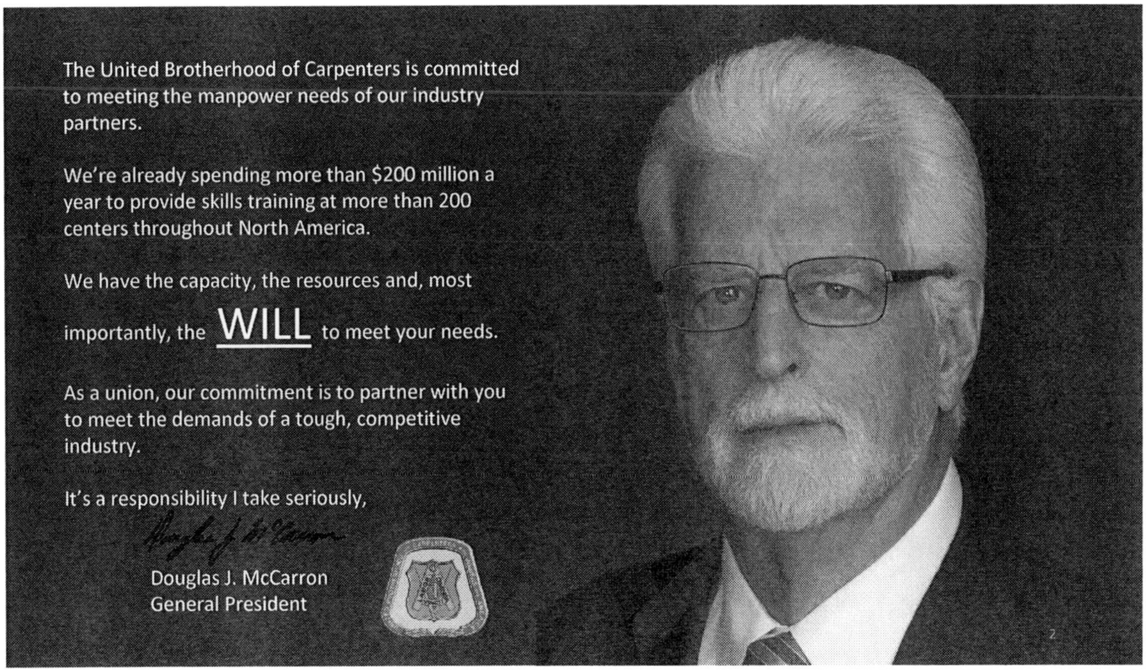
The United Brotherhood of Carpenters is committed to meeting the manpower needs of our industry partners.
We're already spending more than $200 million a year to provide skills training at more than 200 centers throughout North America.
We have the capacity, the resources and, most importantly, the WILL to meet your needs.
As a union, our commitment is to partner with you to meet the demands of a tough, competitive industry.
It's a responsibility I take seriously,
Douglas J. McCarron
General President
2

The mission of the Carpenters International Training Fund (CITF) is to provide state-of-the-art training for members of the United Brotherhood of Carpenters and Joiners of America (UBC).

TRAINING BY THE NUMBERS

- $200 Million+ UBC Skills Training Budget
- 200+ Affiliated training centers
- 6 M+ sq. ft. across North America
- CITF provides airfare, room, and meals

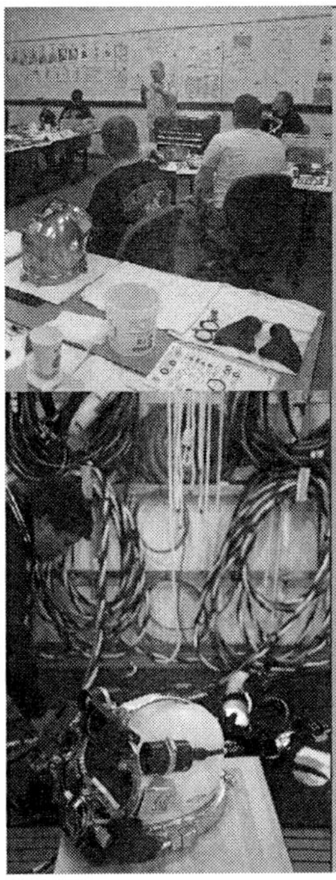

TRAINING BY THE NUMBERS

- 4,450+ Full and part-time instructors
- 100+ Annual TTT courses
- 300+ Courses offered each year
- 1.2 M sq. ft.
- 29 Acres
- 70+ Classrooms

Superintendent CAREER Training

- ***Sponsored by contractor***
- ***Four 3-day workshops over 18 months***
- ***Company mentor***
- ***Transportation, room and meals are furnished by the CITF for the mentor and the trainee***

Stepping up to UBC Foreman

- Foreman essential qualities and skills
- Communications skills
- Safety responsibilities
- Using construction documents
- Role in project meetings
- Running the project
- Achieving the best results
- Utilizing technology

CONTACT YOUR LOCAL TRAINING CENTER FOR MORE INFORMATION

Certified Rigger and Signaler (CRS)

- CRS is a competent professional
- Accredited by NCCA
- Written and Practical Exams
- Exceeds OSHA, ANSI, and ASME standards
- Recognized in U.S., Canada, and Europe

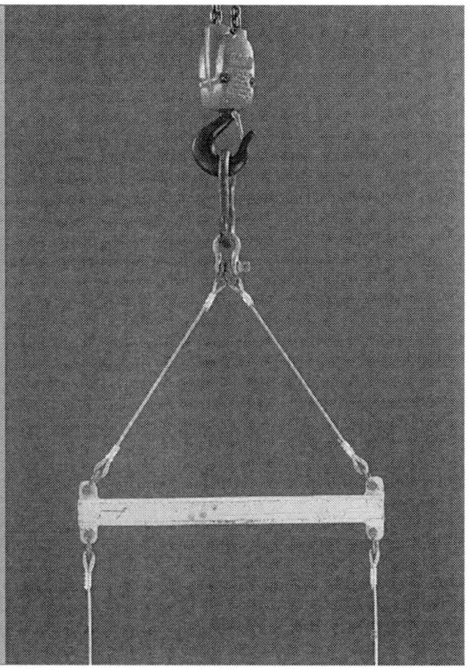

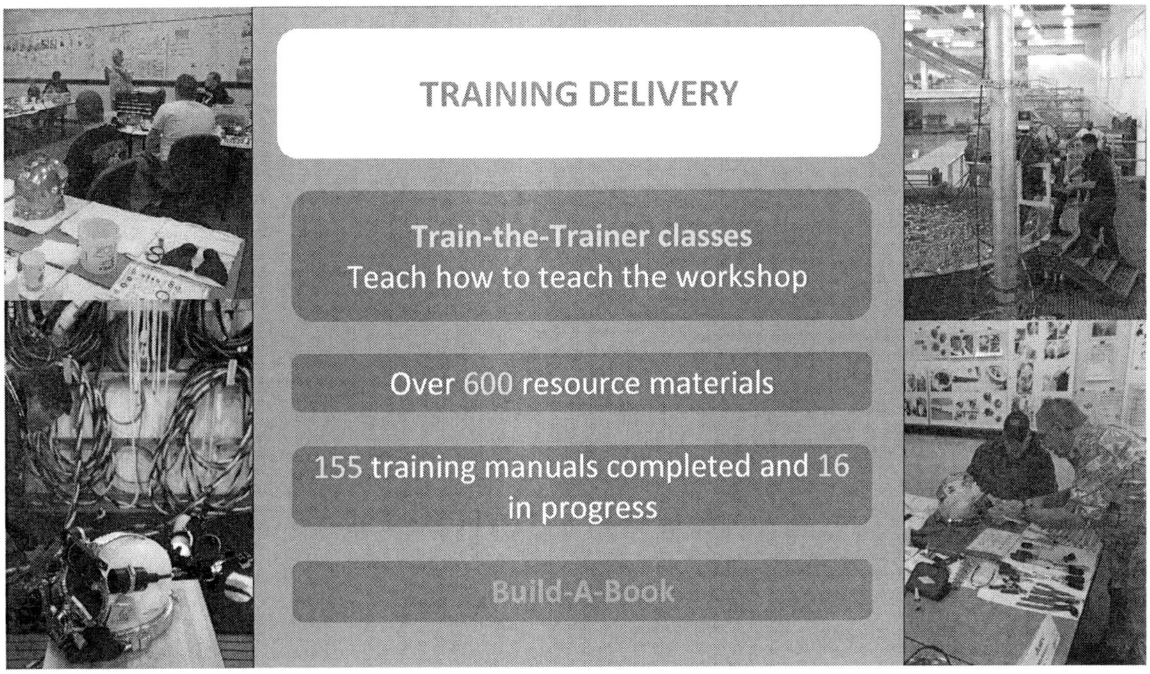
TRAINING DELIVERY
Train-the-Trainer classes
Teach how to teach the workshop
Over 600 resource materials
155 training manuals completed and 16 in progress
Build-A-Book

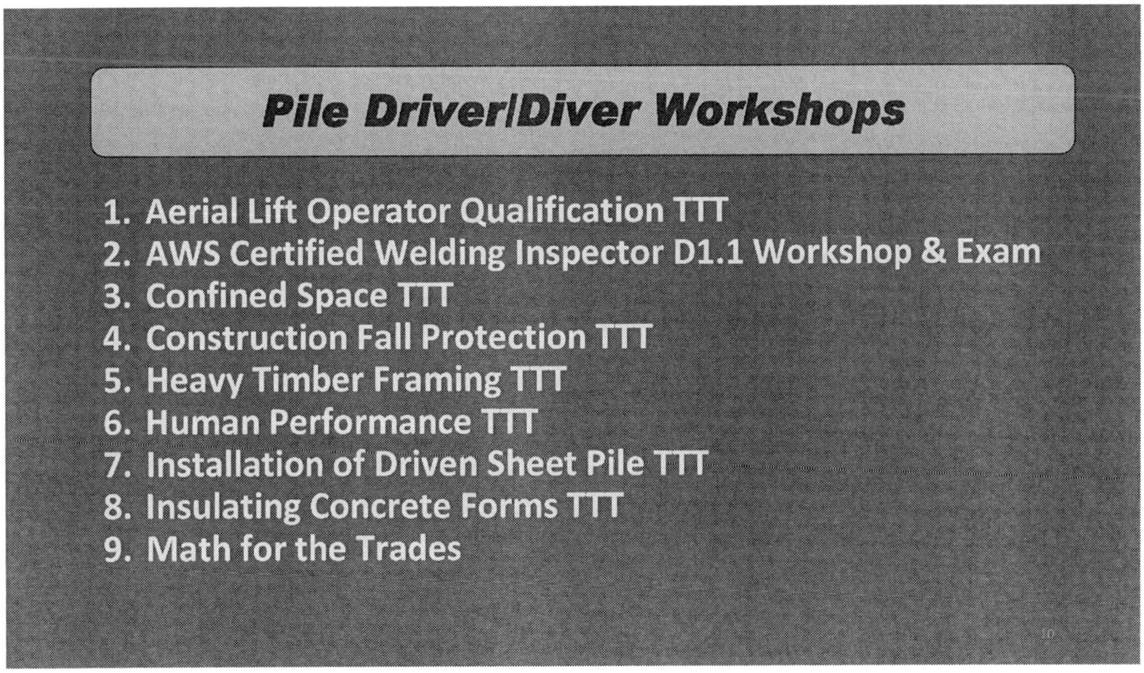
Pile Driver/Diver Workshops
1. Aerial Lift Operator Qualification TTT
2. AWS Certified Welding Inspector D1.1 Workshop & Exam
3. Confined Space TTT
4. Construction Fall Protection TTT
5. Heavy Timber Framing TTT
6. Human Performance TTT
7. Installation of Driven Sheet Pile TTT
8. Insulating Concrete Forms TTT
9. Math for the Trades

Pile Driver/Diver Workshops

10. OSHA 500 Construction TTT
11. OSHA 501 General Industry TTT
12. OSHA 502 Construction TTT
13. OSHA 510 Construction
14. Powered Industrial Truck Operator TTT
15. Print Reading TTT
16. Rigging and Signaling Certification TTT (CRS)
17. Rigging and Signaling Certification TTT Refresher (CRS)
18. Practical Test Proctor Training (CRS)

11

Pile Driver/Diver Workshops

19. Scaffold Erector Qualification TTT
20. Scaffold Erector Qualification TTT Refresher
21. Scaffolding Qualification – Industrial TTT
22. Total Station TTT
23. Total Station 2 TTT
24. Workforce Education 105 - Integrated Curriculum Implementation
25. Workforce Education 106 - Program and Facilities Management
26. Workforce Education 207 - Assessment Techniques
27. Workforce Education 270 – Introduction to Industrial Training

12

Pile Driver/Diver Workshops

28. Workforce Education 413 – Vocational Education for Special Needs Learners
29. Workforce Education 445 – Vocational Guidance
30. Workforce Education 471 – Training in Business and Industry
31. Workforce Education CAS 100A Effective Speech
32. WF ED 310 – Leadership Competencies for Supervisors
33. WF ED 410 – Leadership Competencies for Professionals
34. WF ED 428 – Fiscal Accounting and Management for Workforce Development Professionals
35. WF ED ENGL 015 – Rhetoric and Composition

CITF MEMBER TRAINING

- **Dive Tank and Platform**
- Fiberglass with rubber matts
- 16,000 Gallons Fresh Water
- 12' Deep – 15" Diameter
- View Ports
- Full Complement of Dive Gear
- Heated
- Oil-less Diver Air Compressor
- Overhead Davitt
- Round Robin Communications
- We are not a Dive School
- Diver enhancement courses only

CITF MEMBER TRAINING

- **Underwater Welding Certification**
  Approximately 40 Divers Certified
- **Underwater Bridge Inspection**
  Approximately 120 Divers certified
- **Kirby Morgan Hat Technician**
  Approximately 250 Divers Certified

CITF MEMBER TRAINING

- **Kirby Morgan Hat Technician**

- *Taught by John Ritter*
- *Steve Kushner assistant instructor*
- *Hosted at ITC*
- *Dive Lab certificate*
- *We provide a soft goods kit for your KM helmet*
- *16 hour course*
- *3 year certification*

CITF MEMBER TRAINING
•Underwater Welding Certification
•We work with AWS SCWI to produce PQR's
•AWS D3.6M compliant qualifications
•Several WPS's to qualify to
•In house CWI for immediate qualification

CITF MEMBER TRAINING
•HYTROC - Underwater Hydraulic Tensioning
•Underwater Bridge Inspection
•NHI Course
•Taught by Collins Engineering
•Hosted at ITC
•32 hour lecture type course

What's on my Training Verification Card?

- ***Current membership information***
- ***Training history***
- ***Welding and Diving certifications***
- ***Additional Qualifications and Certifications***
- ***Upcoming classes***
- ***LMS updates/notices***

Training Verification Card (TVC)

- Unique QR Code For Each Member
- https://train.carpenters.org/tvc?x=bD1YyOPU4T

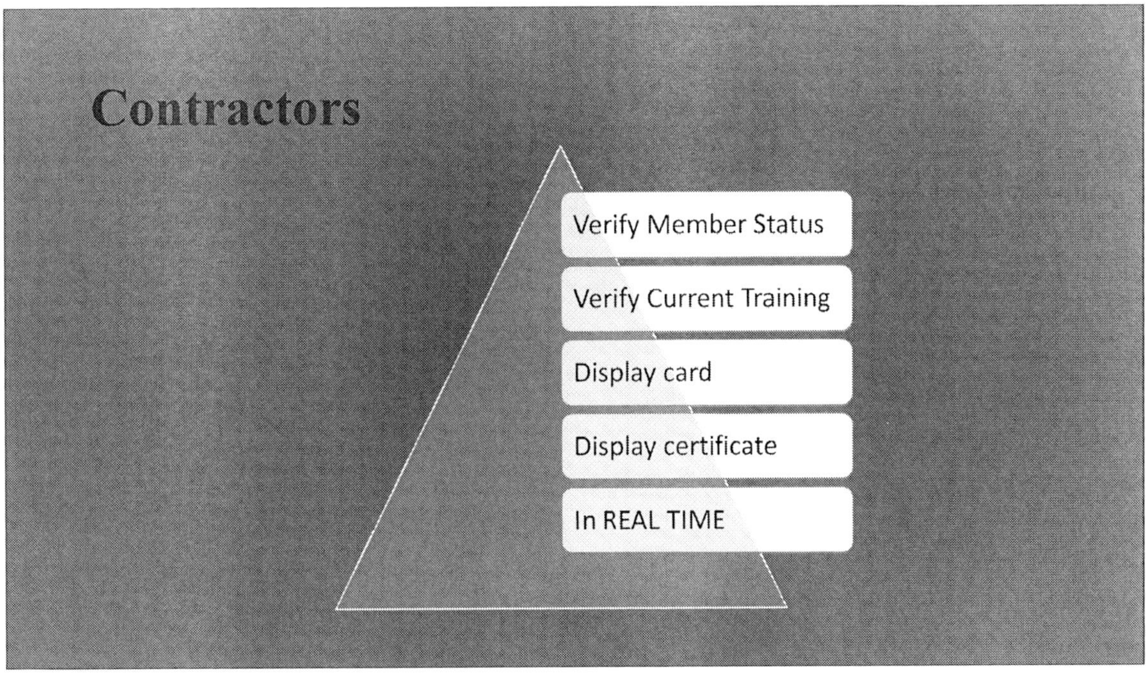
Contractors
Verify Member Status
Verify Current Training
Display card
Display certificate
In REAL TIME

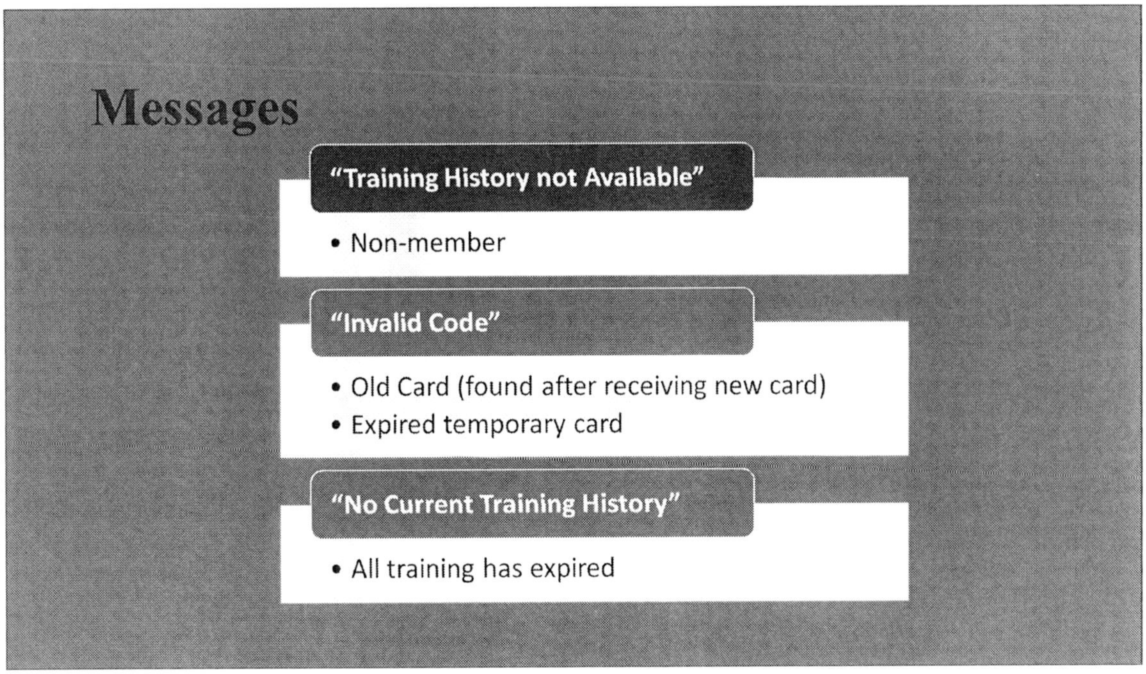
Messages
"Training History not Available"
• Non-member
"Invalid Code"
• Old Card (found after receiving new card)
• Expired temporary card
"No Current Training History"
• All training has expired

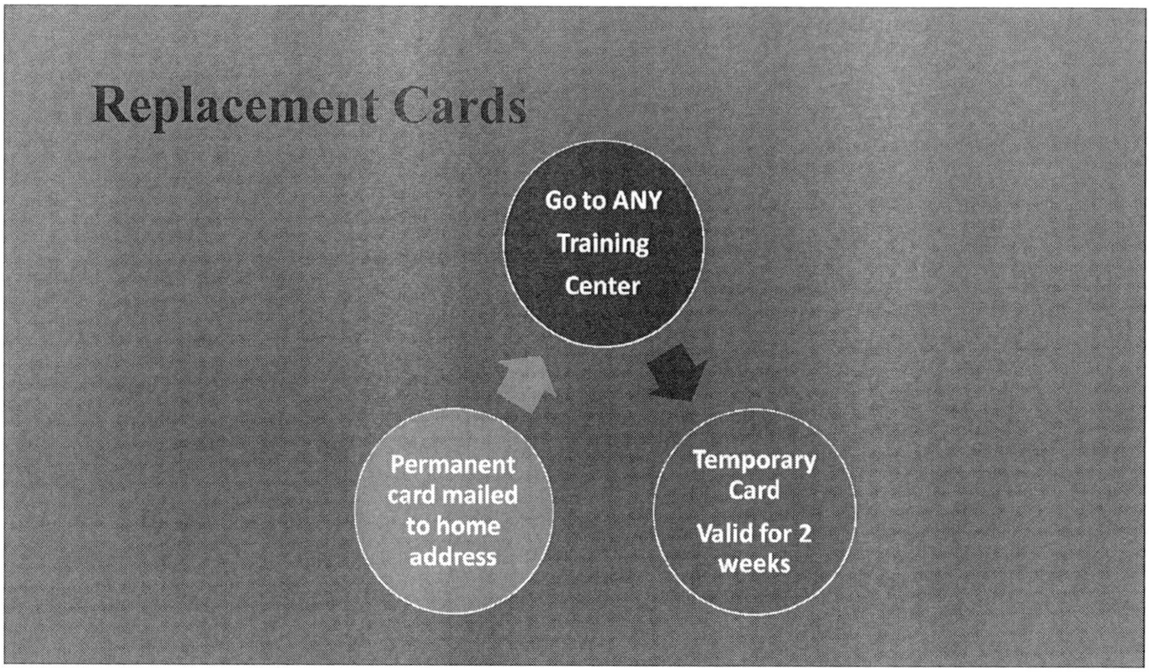
Replacement Cards
Go to ANY Training Center
Temporary Card
Valid for 2 weeks
Permanent card mailed to home address

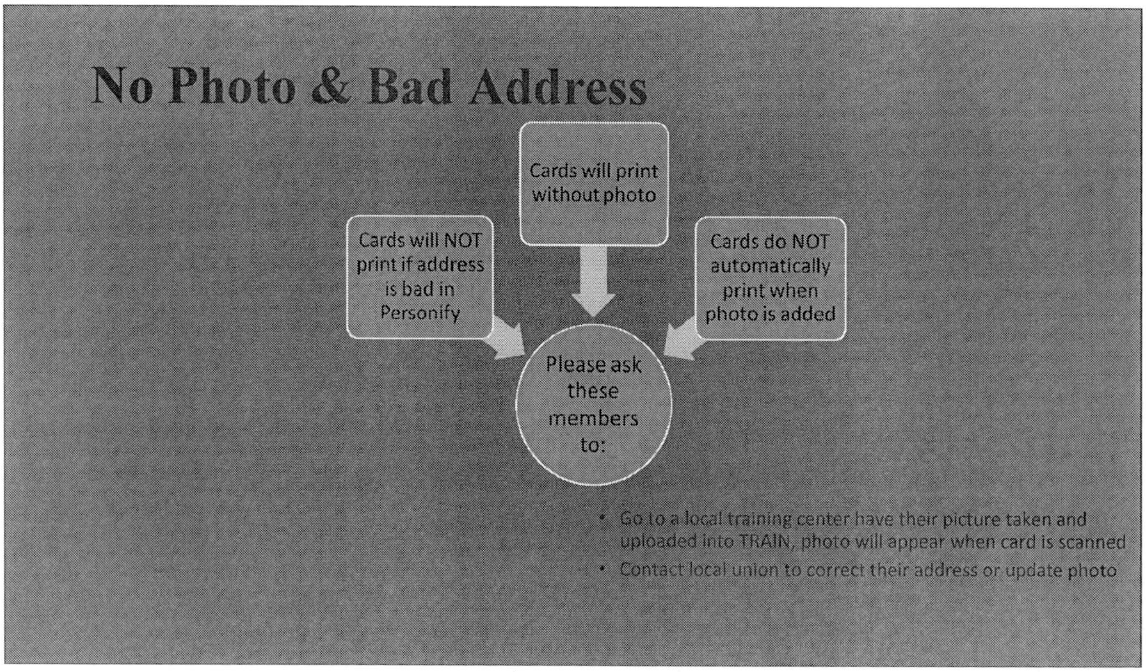
No Photo & Bad Address
Cards will print without photo
Cards will NOT print if address is bad in Personify
Cards do NOT automatically print when photo is added
Please ask these members to:
• Go to a local training center have their picture taken and uploaded into TRAIN, photo will appear when card is scanned
• Contact local union to correct their address or update photo

Mailing

TVCs	• Mailed to the member's home address
Pending Apprentices	• Will receive card when they become members
Previous Members	• Their previous card will reactivate, or if they never received a card, it will print and be mailed

Scanned QR Code

- Use a QR code reader application on a smartphone to view the records
- Send link to PC or
- Use PC with QR code reader software and webcam

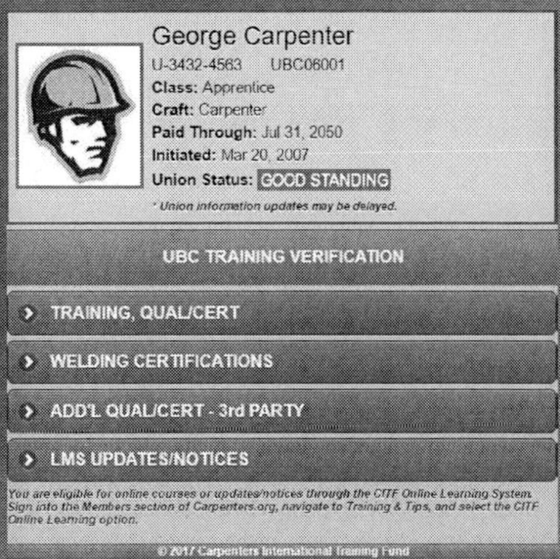

Training, Qual/Cert

- Displays current training

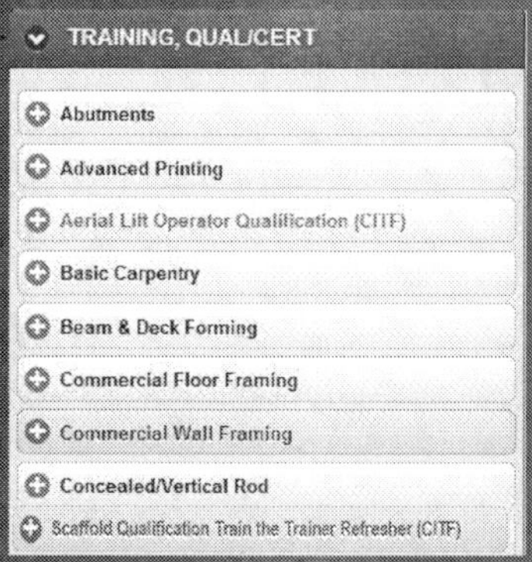

- Expired and pending will NOT display
- Training expiring within 60 days is
- Automatically updates every time a member completes a class, certification, or qualification

Training, Qual/Cert

- Drill down
  - Date of training or expiration date
  - Cards and Certificates

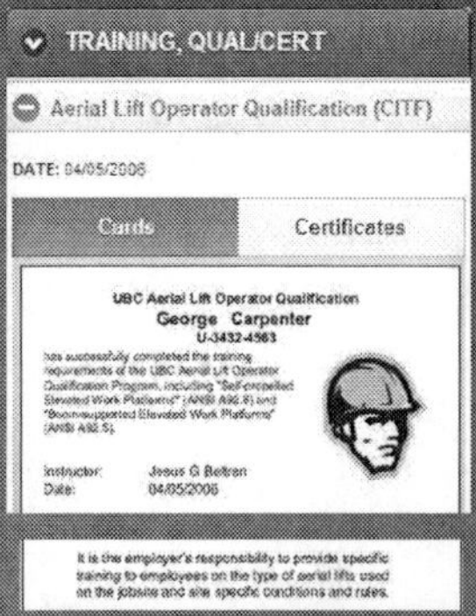

Welding Certifications

- Drill down to see list
- Drill down for details
  - Viewable PDF

Additional Qual/Certs – 3rd Party

- Qual/Cert Number can include description
- Viewable PDF

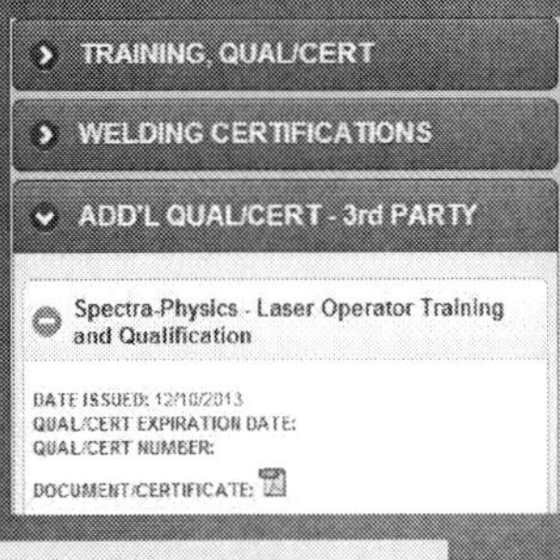

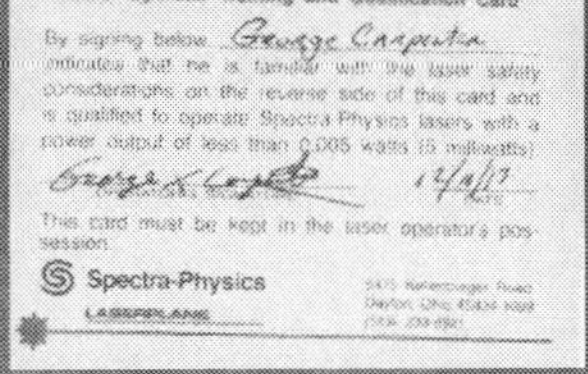

Laser Operator Training and Qualification Card

By signing below George Carpenter indicates that he is familiar with the laser safety considerations on the reverse side of this card and is qualified to operate Spectra-Physics lasers with a power output of less than 0.005 watts (5 milliwatts)

This card must be kept in the laser operator's possession.

Spectra-Physics

LASERPLANE

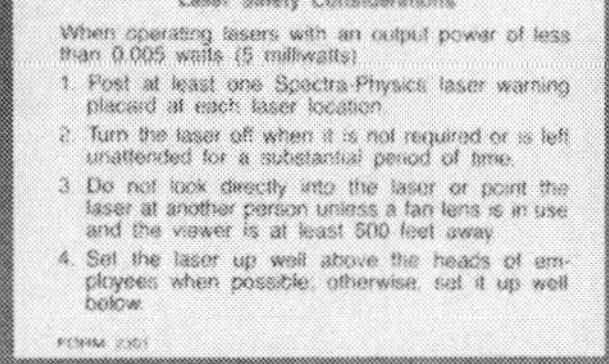

Laser Safety Considerations

When operating lasers with an output power of less than 0.005 watts (5 milliwatts)

1. Post at least one Spectra-Physics laser warning placard at each laser location.
2. Turn the laser off when it is not required or is left unattended for a substantial period of time.
3. Do not look directly into the laser or point the laser at another person unless a fan lens is in use and the viewer is at least 500 feet away.
4. Set the laser up well above the heads of employees when possible; otherwise, set it up well below.

FORM 2301

LMS Updates/notices

- Drill down to see list
- Drill down for date item completed

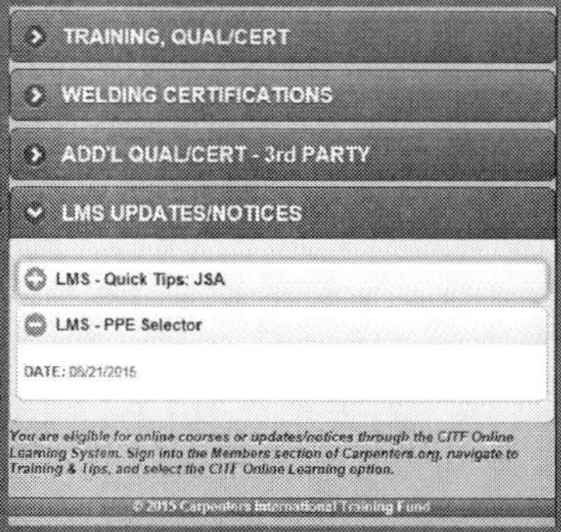

Order Certificates & TVC View TVC

- Members still receive certificates for every CITF class, and all certifications and qualifications
- Local training centers can order replacement certificates and new TVC cards in TRAIN

Local Centers can also view TVC in TRAIN

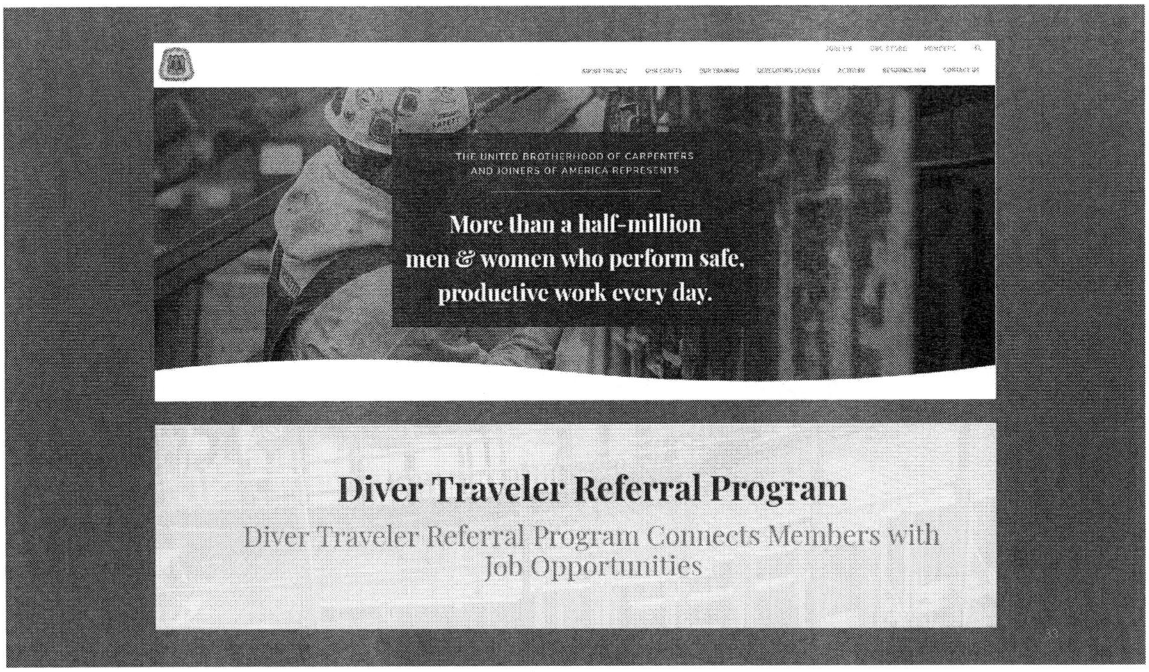
THE UNITED BROTHERHOOD OF CARPENTERS
AND JOINERS OF AMERICA REPRESENTS
More than a half-million
men & women who perform safe,
productive work every day.
Diver Traveler Referral Program
Diver Traveler Referral Program Connects Members with Job Opportunities

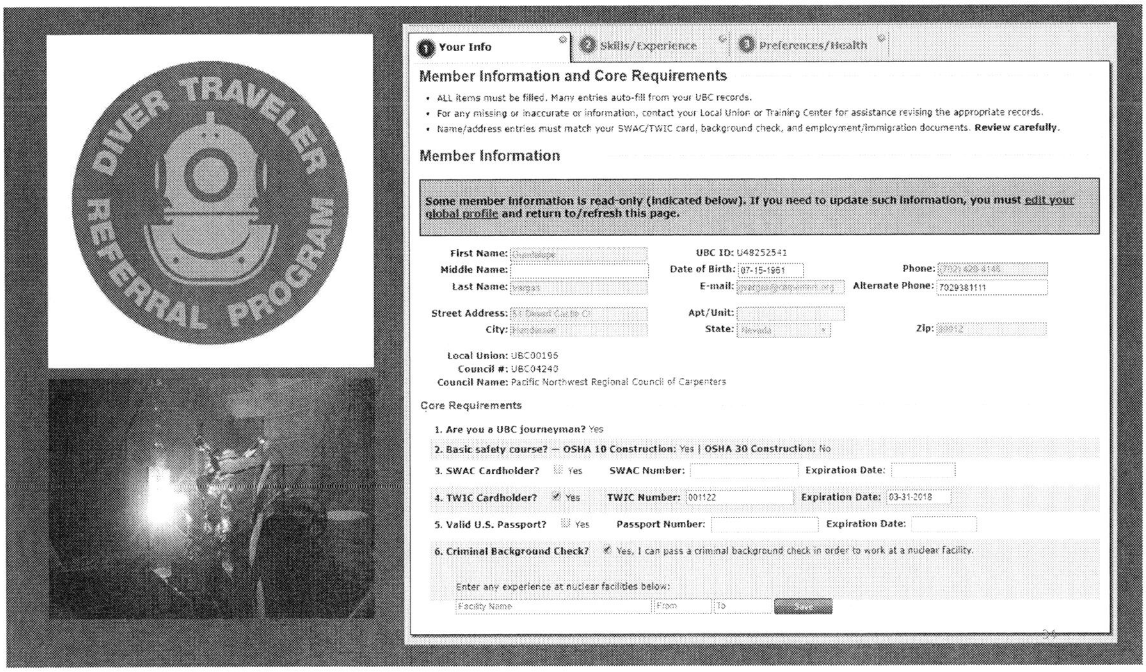
DIVER TRAVELER
REFERRAL PROGRAM
Your Info
Skills/Experience
Preferences/Health
Member Information and Core Requirements
ALL items must be filled. Many entries auto-fill from your UBC records.
For any missing or inaccurate or information, contact your Local Union or Training Center for assistance revising the appropriate records.
Name/address entries must match your SWAC/TWIC card, background check, and employment/immigration documents. Review carefully.
Member Information
Some member information is read-only (indicated below). If you need to update such information, you must edit your global profile and return to/refresh this page.
First Name:
UBC ID: U48252541
Middle Name:
Date of Birth: 07-15-1961
Phone:
Last Name:
E-mail:
Alternate Phone: 7029381111
Street Address:
Apt/Unit:
City:
State:
Zip:
Local Union: UBC00196
Council #: UBC04240
Council Name: Pacific Northwest Regional Council of Carpenters
Core Requirements
1. Are you a UBC journeyman? Yes
2. Basic safety course? — OSHA 10 Construction: Yes | OSHA 30 Construction: No
3. SWAC Cardholder? Yes SWAC Number: Expiration Date:
4. TWIC Cardholder? Yes TWIC Number: 001122 Expiration Date: 03-31-2018
5. Valid U.S. Passport? Yes Passport Number: Expiration Date:
6. Criminal Background Check? Yes, I can pass a criminal background check in order to work at a nuclear facility.
Enter any experience at nuclear facilities below:
Facility Name
From
To

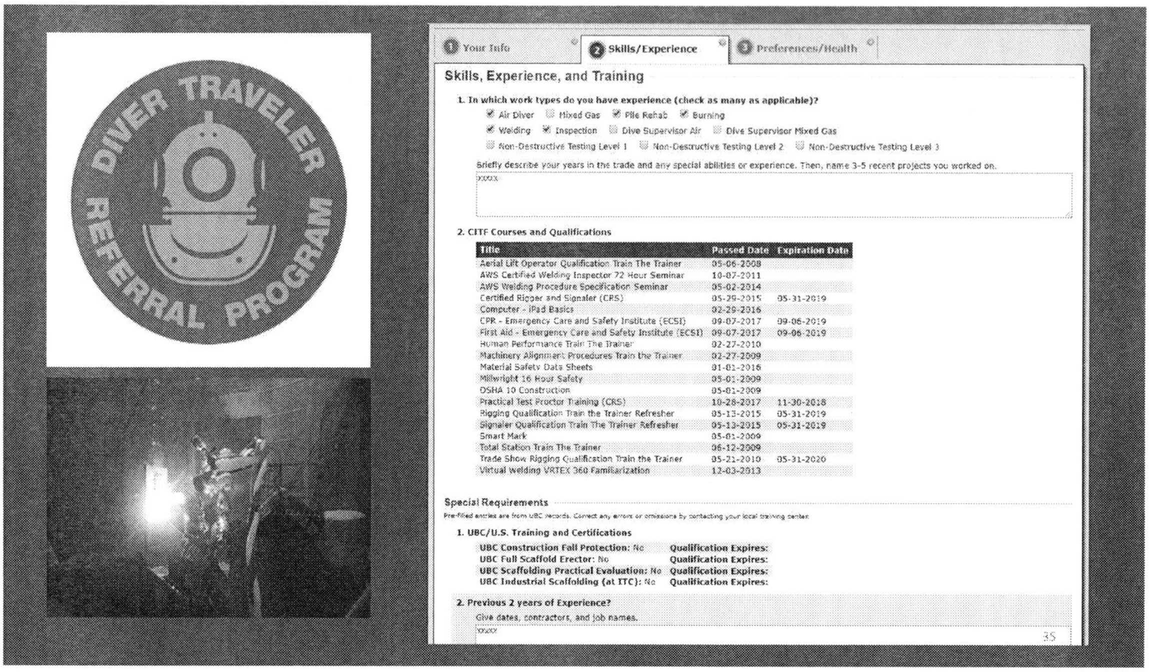

Your Info | Skills/Experience | Preferences/Health

Skills, Experience, and Training

1. In which work types do you have experience (check as many as applicable)?

- [x] Air Diver
- [ ] Mixed Gas
- [x] Pile Rehab
- [x] Burning
- [x] Welding
- [x] Inspection
- [ ] Dive Supervisor Air
- [ ] Dive Supervisor Mixed Gas
- [ ] Non-Destructive Testing Level 1
- [ ] Non-Destructive Testing Level 2
- [ ] Non-Destructive Testing Level 3

Briefly describe your years in the trade and any special abilities or experience. Then, name 3-5 recent projects you worked on.

xxxx

2. CITF Courses and Qualifications

Title	Passed Date	Expiration Date
Aerial Lift Operator Qualification Train The Trainer	05-06-2008	
AWS Certified Welding Inspector 72 Hour Seminar	10-07-2011	
AWS Welding Procedure Specification Seminar	05-02-2014	
Certified Rigger and Signaler (CRS)	05-29-2015	05-31-2019
Computer - iPad Basics	02-29-2016	
CPR - Emergency Care and Safety Institute (ECSI)	09-07-2017	09-06-2019
First Aid - Emergency Care and Safety Institute (ECSI)	09-07-2017	09-06-2019
Human Performance Train The Trainer	02-27-2010	
Machinery Alignment Procedures Train the Trainer	02-27-2009	
Material Safety Data Sheets	01-01-2016	
Millwright 16 Hour Safety	05-01-2009	
OSHA 10 Construction	05-01-2009	
Practical Test Proctor Training (CRS)	10-28-2017	11-30-2018
Rigging Qualification Train the Trainer Refresher	05-13-2015	05-31-2019
Signaler Qualification Train The Trainer Refresher	05-13-2015	05-31-2019
Smart Mark	05-01-2009	
Total Station Train The Trainer	06-12-2009	
Trade Show Rigging Qualification Train the Trainer	05-21-2010	05-31-2020
Virtual Welding VRTEX 360 Familiarization	12-03-2013	

Special Requirements

Pre-filled entries are from UBC records. Correct any errors or omissions by contacting your local training center.

1. UBC/U.S. Training and Certifications

UBC Construction Fall Protection: No **Qualification Expires:**
UBC Full Scaffold Erector: No **Qualification Expires:**
UBC Scaffolding Practical Evaluation: No **Qualification Expires:**
UBC Industrial Scaffolding (at ITC): No **Qualification Expires:**

2. Previous 2 years of Experience?

Give dates, contractors, and job names.

xxxx

35

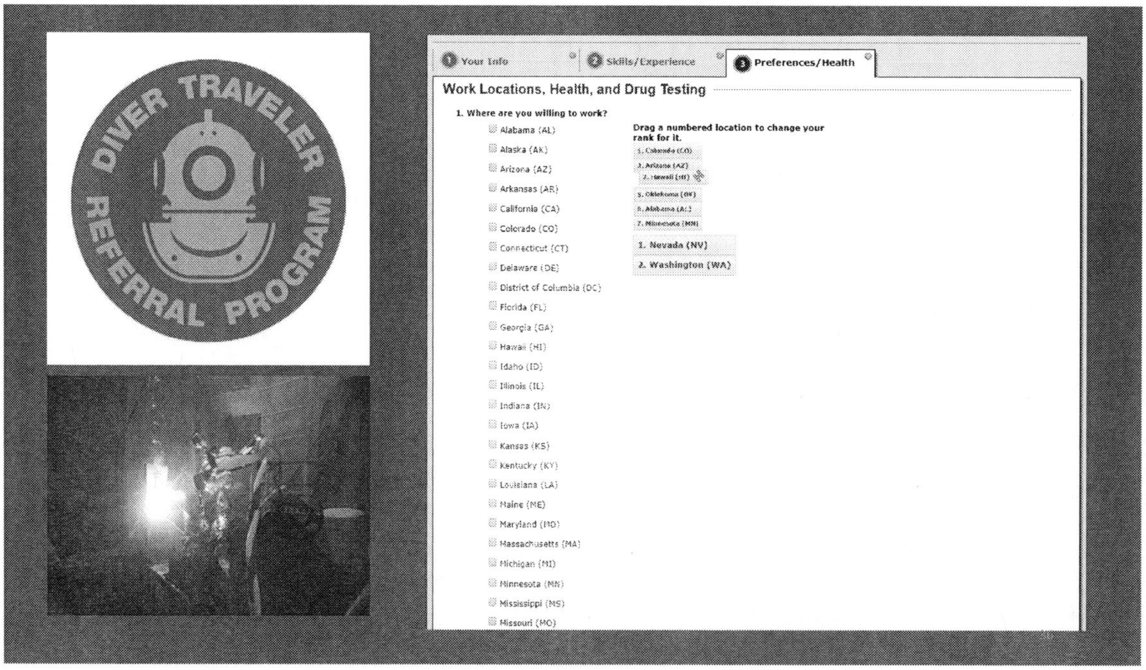

Your Info | Skills/Experience | Preferences/Health

Work Locations, Health, and Drug Testing

1. Where are you willing to work?

- [ ] Alabama (AL)
- [ ] Alaska (AK)
- [ ] Arizona (AZ)
- [ ] Arkansas (AR)
- [ ] California (CA)
- [ ] Colorado (CO)
- [ ] Connecticut (CT)
- [ ] Delaware (DE)
- [ ] District of Columbia (DC)
- [ ] Florida (FL)
- [ ] Georgia (GA)
- [ ] Hawaii (HI)
- [ ] Idaho (ID)
- [ ] Illinois (IL)
- [ ] Indiana (IN)
- [ ] Iowa (IA)
- [ ] Kansas (KS)
- [ ] Kentucky (KY)
- [ ] Louisiana (LA)
- [ ] Maine (ME)
- [ ] Maryland (MD)
- [ ] Massachusetts (MA)
- [ ] Michigan (MI)
- [ ] Minnesota (MN)
- [ ] Mississippi (MS)
- [ ] Missouri (MO)

Drag a numbered location to change your rank for it.

1. Nevada (NV)
2. Washington (WA)

IMPORTANT THINGS TO REMEMBER:

- This online registration system is the only way to get into the UBC Diver TRP.
- You must have all your certifications and qualifications entered into the TRAIN system.
- No one can predict when or if you will get contacted to travel. There is no "average wait" or "probably by" date.

Certified divers who are willing to travel should register with this system and have their information available to dispatchers across the United States.

Check out this link to get started! – UBC Diver Traveler Referral Program

37

UI2018
UNDERWATER
INTERVENTION

New Orleans, LA
February 6-8, 2018

Gateway applications of Marine Robotics in the offshore Underwater energy sector

Sudhir Pai
Managing Director
Schlumberger Robotics Services

pai2@slb.com
www.slb.com/robotics

Reduce divers in the water and humans at sea
Protect assets
Without compromising data quality

Schlumberger

Business Need for Autonomous Vehicles

- Data in difficult to access environments
- Environmental understanding
- Environmental compliance
- Asset integrity

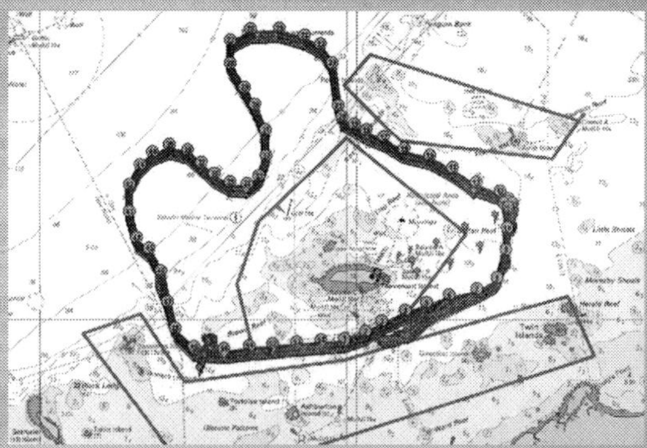

Schlumberger

Technology and Business Applications

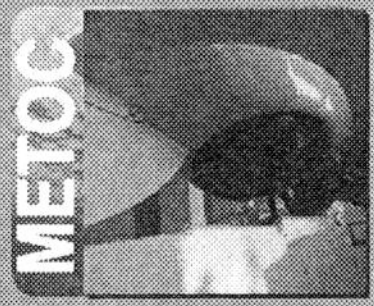

Metrology & Oceanography

Hydrocarbon Monitoring

Subsea Data Communications

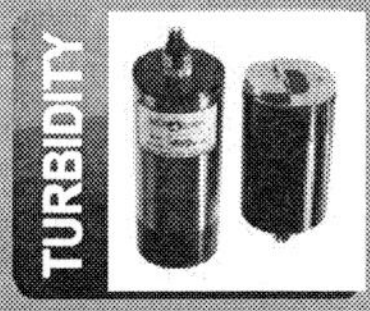

Water Quality & Particle Suspension

Complement Towed Streamer Surveys

Offshore Magnetic Measurements

Marine mammal Monitoring & Fisheries

Schlumberger

3

Deployment Techniques used in Oil and Gas

- Drill Pipe
- Wireline
- Towed Streamer Surveys
- Autonomous Vehicles
  - Surface
    - Wave propulsion systems
    - Wind
    - Diesel; Battery
  - Underwater: ROV, AUV
  - Ariel: Drones

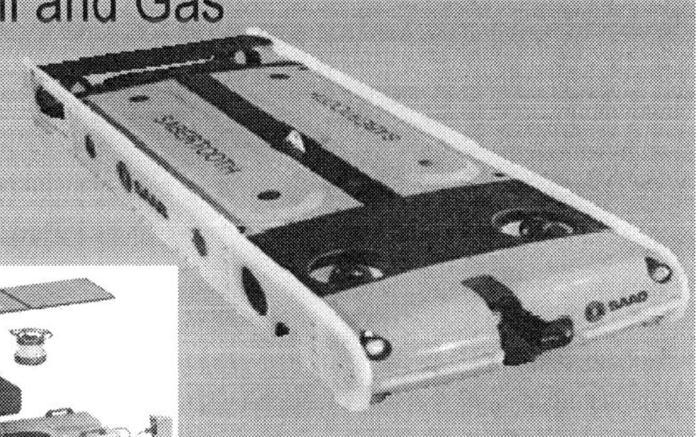

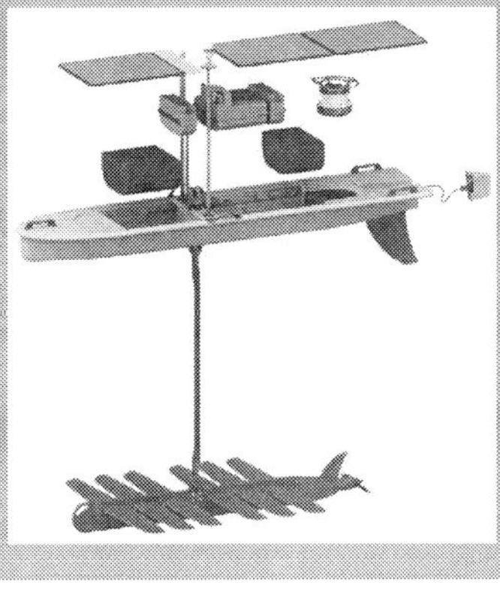

Schlumberger

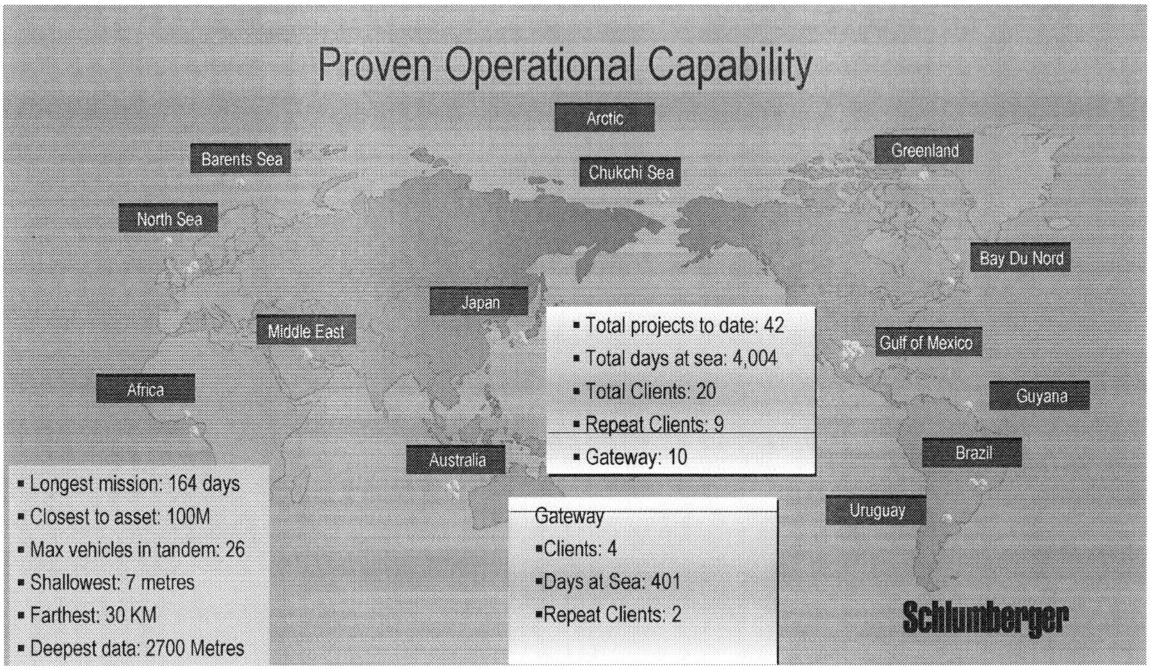

Gateway Operations – 5G and 6G

Data Gateway: surface communications with subsea equipment via acoustic modem and transfer of data onwards via satellite.

- Shore piloted acquisition from sub-sea acoustic modem to client office
- Sonardyne 5G and 6G acoustic comms
- Iridium RUDICS satellite comms
- Surface data path doesn't have to be satellite, it could be radio or cellular
- Low HSE exposure and high efficiency compared to practice of harvesting data by support vessel with dunker
- Harvest one or many bottom systems
- Single AMV can provide multiple harvests without multiple deployments (holding pattern while standing by)
- Depths to 5000m+

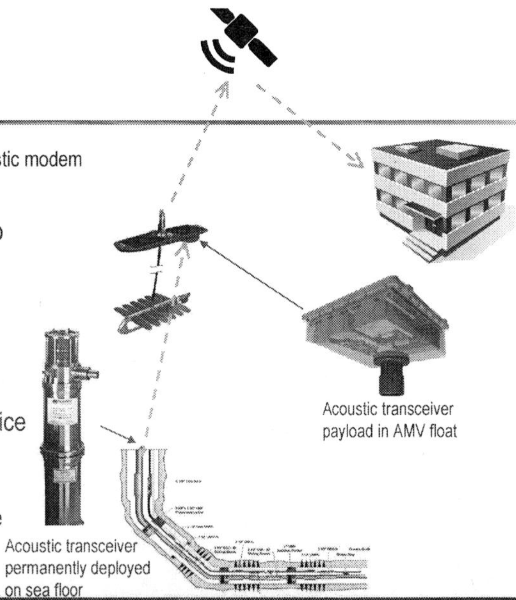

Schlumberger

Case Study: DeepWater GoM Gateway

Shell

For the last three years, an array of Sonardyne's autonomous sensors has been recording pressure data for monitoring seafloor subsidence in a Shell-operated deepwater field.

A Wave Glider fitted with a Sonardyne high speed acoustic modem has completed multiple data harvesting operations and retrieved logged data on demand without the intervention of a conventional survey vessel.

"Using a Wave Glider to harvest data acoustically from seabed instruments is a highly economical, and sustainable way to monitor our production," said Paul Hatchell, Geophysicist in Shell's Areal Monitoring research team.

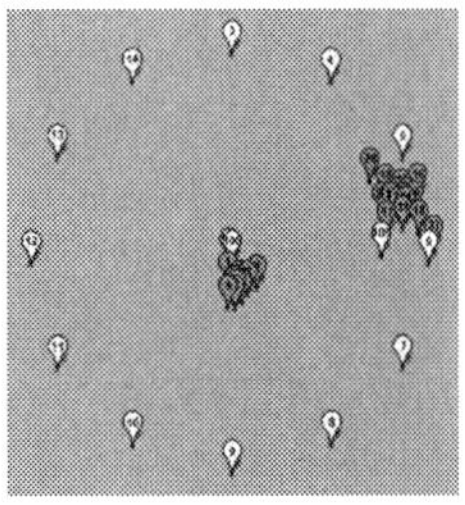

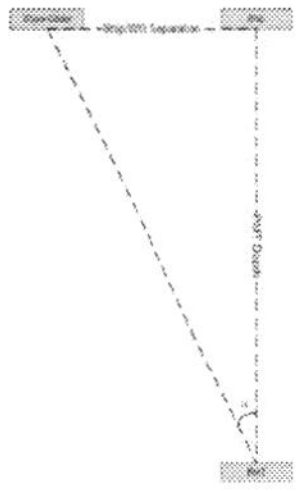

View of clusters of PMT locations

- Pressure monitoring
- Data harvesting on demand
- Wave Glider replacing vessels

18th Annual Gulf of Mexico Deepwater Technical Symposium

Schlumberger

7

Gateway Operations – Tullow - Ghana

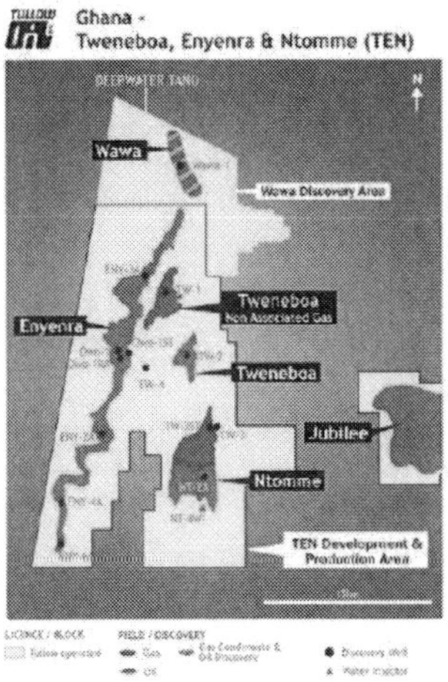

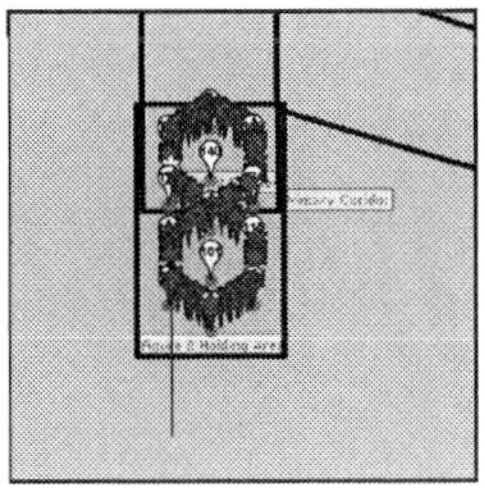

CHALLENGE

Improve access to important well information from subsea tree-mounted acoustic data loggers (ADLs) used during a well testing program in 1,200-m water depth.

SOLUTION

Deploy an autonomous marine vehicle equipped with an acoustic modem to harvest downhole pressure and temperature (DHPT) measurements collected by the ADLs for subsequent recording, transfer via satellite to a secure server, and analysis by Tullow Oil's London-based reservoir team.

RESULTS

Better informed decision making and decreased data retrieval costs through autonomous harvesting by an unmanned surface vehicle (USV) of raw data for automatic transmission by satellite to the reservoir team for analysis.

"Data acquired using the Wave Glider provided valuable insight into reservoir communication pathways between development wells, and in turn enabled robust planning for the start-up phase of the TEN Project. This was all achieved at low cost when compared to standard industry practices and reduced exposure of manpower and support vessel resources in the field. As with all projects, people interfaces and communication lines are key to success—it was a pleasure working with all of the contributors to the highly successful Wave Glider project and we look forward to looking for additional opportunities to apply this technology again."

Robin Rindfuss
TEN Project, Subsurface Manager
Tullow Oil

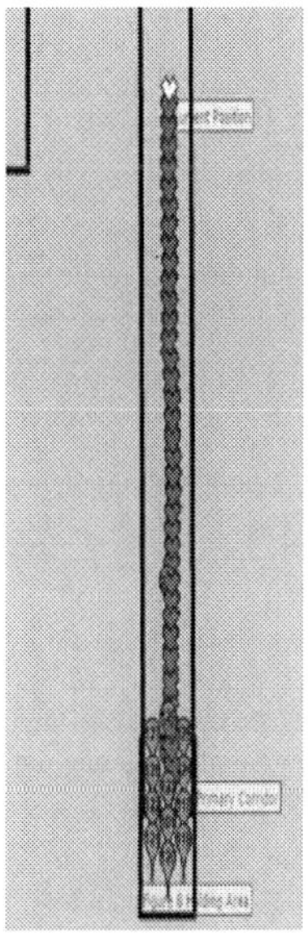

Route heading north to a wellhead location as shown by the GPS updates (red x's).

ADL Transit

Gateway Operations – Jogmec – Japan 2016-18

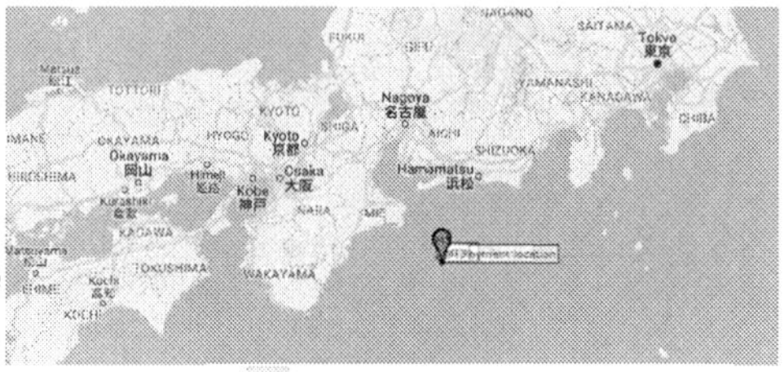

- SV2 AMV, 6G Data Harvest, ~1000m water depth
- AMV experiences drift due to high currents and low sea state
- Successful data harvest during 3 drift passes

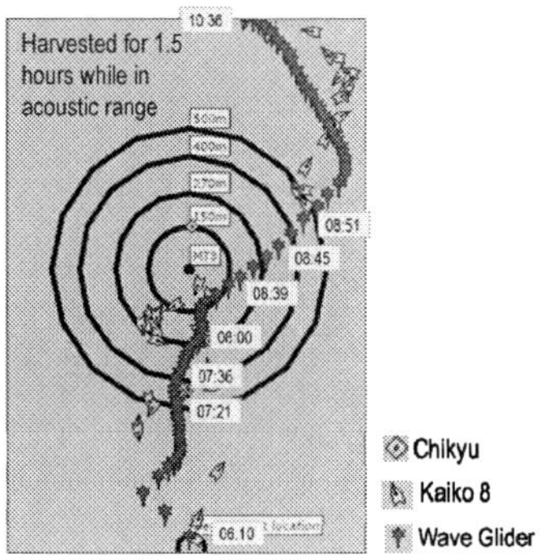

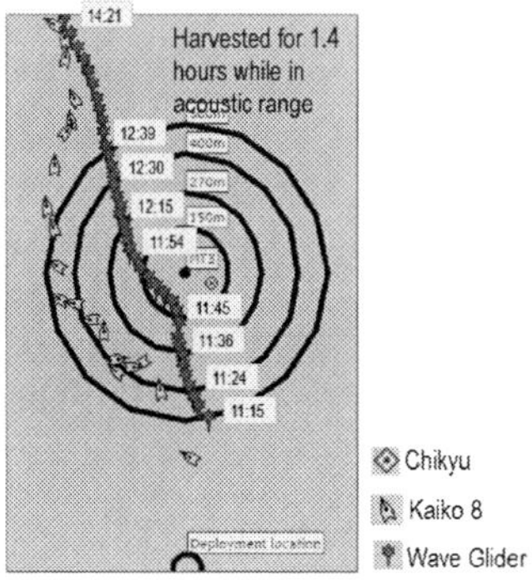

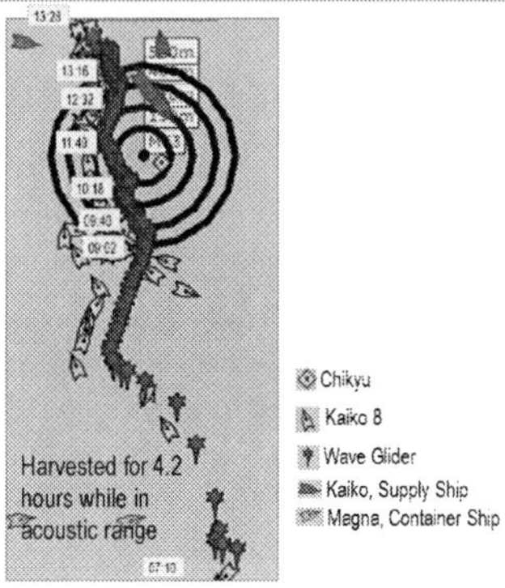

2017 Commercial IMR Offering

AUV Inspection Service

Equipment

600m (2,000ft) rated, Kongsberg MUNIN Torpedo AUV rental

1,200m (4,000ft) rated, SAAB Sabertooth Hovering AUV rental

Kongsberg MUNIN

Marine Support

Vessels of opportunity

Client chartered vessels

Applications

Pipeline Inspection

Route/ Site Surveys

Seabed Mapping

Residency FEED Studies

- Vehicle Agnostic
- Field Specific Integration

SAAB Falcon Seaeye

OPTIMIZE FROM PORE TO PROCESS

2018 Commercial IMR Offering

AUV Inspection Service

Equipment

•3,000m (10,000 ft) rated, SAAB Sabertooth Hovering AUV owned

•3,000m (10,000 ft) rated, SAAB Sabertooth Hovering AUV owned

Marine Support

•Vessels of opportunity

•Client chartered vessels

Applications

•Pipeline Inspection

•Route/ Site Surveys

•Seabed Mapping

Residency Development

- Docking Station

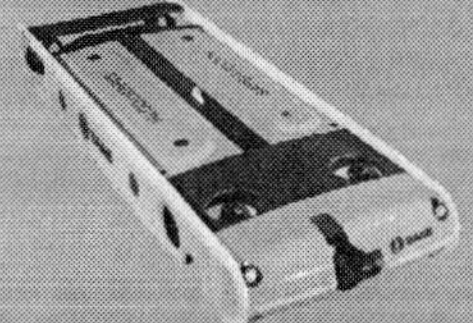

SAAB Sabertooth

Schlumberger

Slide 11

Conclusions and results

- HSE: USV demonstrated a safe cost-effective alternative remote data harvesting method
- Cost: USV has demonstrated self-navigation capability to avoid collision when operating nearby a drillship. They replaced vessels at a substantial cost reduction.
- Excellence in Execution: USV has demonstrated its remote navigation capability to hold on station and circling as per defined waypoint patterns as small as 50 m radius circle
- Data Quality: Quality of all harvested data by USV has been validated with no data corruption

Schlumberger

Slide 12

Acknowledgements

- Clients:
  - Shell
  - Tullow
  - Jogmec
- Service Providers:
  - Schlumberger
  - Sonardyne Intl
  - Liquid Robotics Inc, a Boeing Company

UI2018 UNDERWATER INTERVENTION

New Orleans, LA
February 6-8, 2018

Schlumberger

Underwater Examinations for Classification

John Preston | February 7, 2018

ABS

Safety Moment

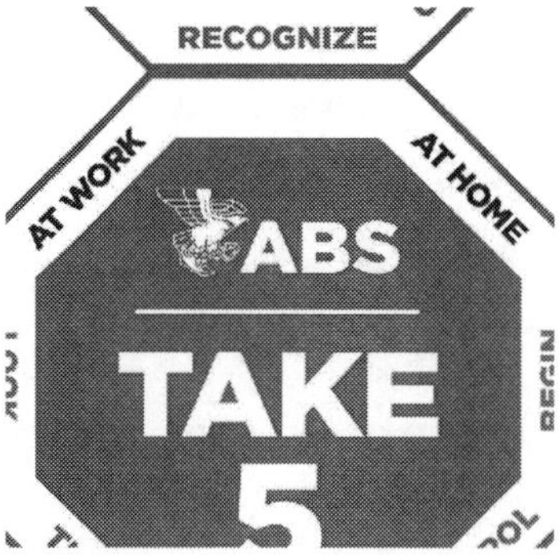

ABS

160 Years of Rules

- Building and Classing Steel Vessels
- Mobile Offshore Drilling Units
- Floating Production Installations
- Offshore Installations
- Single Point Moorings
- Underwater Vehicles, Systems, and Hyperbaric Facilities
- Offshore Support Vessels
- High Speed Craft
- Barges

3

ABS In Water Examination Requirements- Hull

- General examination of 100% of underwater hull
  - Rudders, propellers, propeller shaft clearances
  - Close visual examination or NDE of critical areas depending on inspection plan
    - May require spot cleaning or extensive cleaning
  - Check corrosion protection potentials
    - Anode survey
  - Splash zone examination
  - Thickness readings
- Examine all internal compartments
  - Thickness readings
- Sea valve examination

4

ABS In Water Examination Requirements – Mooring

- Chain visual exam
- Check for chain trenching or fretting
- Chain cleaning and gauging in selected areas
  - In way of grips
  - Length over 5 links
  - Top chain
  - Wasted areas
  - Touchdown areas
  - Splash zone
- Soft and wire rope
- Pile exam
- Visual examination of tendons
  - Flooded member detection
  - Flex elements
- Corrosion protection readings
- Visual exam of anodes
- Support or spring buoys
- Any high fatigue areas

5

ABS In Water Examination Requirements – Pipe

- Riser and pipeline examination
  - Splash zone
  - Crossings and free spans
  - Touchdown area
  - High stress areas
  - Areas of low fatigue life
  - End terminations and high bending areas
  - Previous repairs or modifications
  - Previous anomalies

6

Traditional Underwater Examination Methods

- All parties are at the scene onboard the vessel or unit
- ROVs controlled manually
- General visual examination of all items
- Close up examination of selected items
- Spot cleaning with ROV or divers
- NDE conducted by divers
- Manual chain gauging

7

Leading Edge ROV Use

- ROV tools
  - Eddy current inspection
  - Alternating current field measurement
  - Ultrasonic shear wave, phased array and thickness testing
  - Crawlers
  - Pencil cameras
  - Floating cameras
- Internal ballast tank examinations
  - Gauging with ROV

Photo: Oceaneering

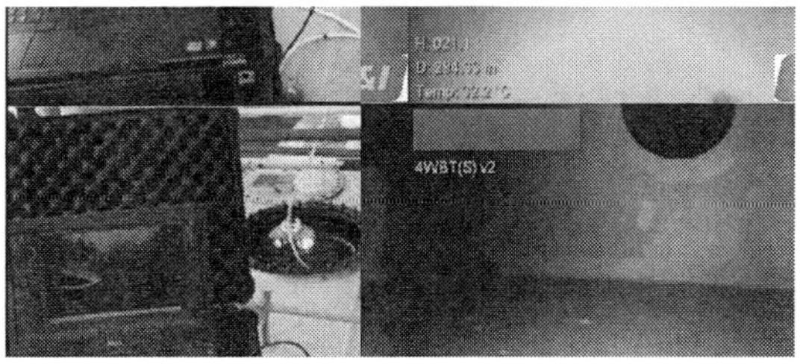

Photo: Oceaneering

8

Leading Edge ROV Use

- Optical examination of chain
  - Grip area thickness
- 3D optical or laser scanning
- Magnetic flux leakage mooring wire examination

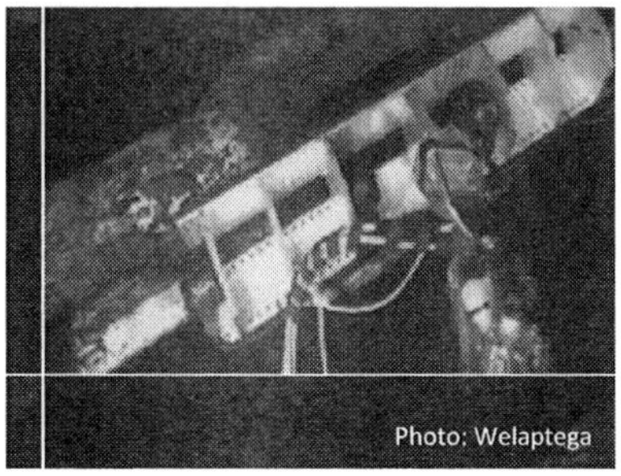

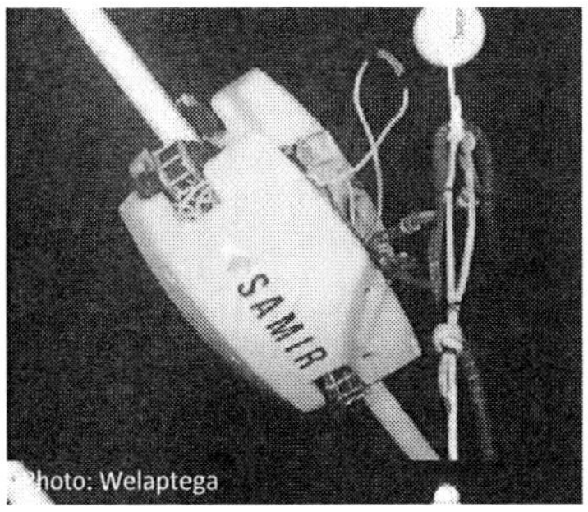

9

Leading Edge ROV Use

- Remote ROV examinations
  - Class and owner witnesses stay on shore
- Specialized tooling
  - Tendon flex element examination
- Sea valve examination
  - Blanking sea chests
  - Internal examinations

10

ABS

Upcoming ROV Methods

- Semi autonomous examination methods
- Autonomous ROV examinations
- 3D Models
  - Laser scan
  - Stitched photos and models
- Automated comparative examinations
  - Compare previous model to current model
- Automated chain cleaning
- Friction stir welding

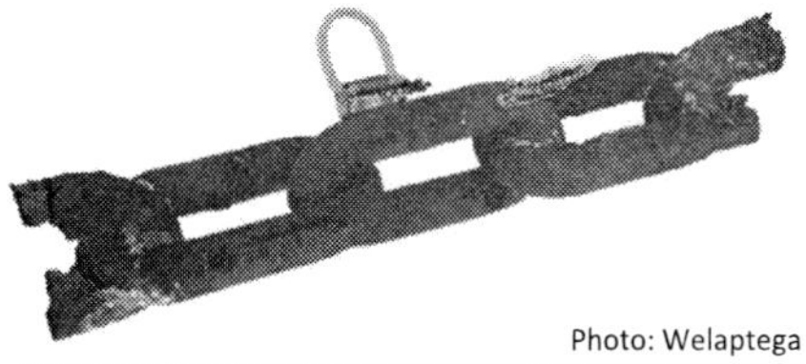

Photo: Welaptega

Photo: Oceaneering

11

ABS

When We Can Have This?

- Autonomous ROV flies the area, laying out its own course
  - Hull
  - Mooring
  - Risers
  - Ballast tanks
  - Piping
- Compares current conditions to previous conditions
- Identifies areas that require cleaning or further examination
- Cleans areas, measures, and evaluates new readings
  - Conducts required NDE
  - Repairs
- Produces simple graphical result or model, identifying amount of change
  - Growth weight
  - Wastage
  - Erosion
  - Physical damage

12

ABS

ABS

Thank You

www.eagle.org

bluefutures

Engaging the Next Generation

Liesl Hotaling, *VP of Education, MTS*

marine technology SOCIETY
Opportunity runs deep

Introduction

- The web of natural and anthropogenic ocean-related issues increases the importance of engaging youth about the ocean and growing list of complex issues.
- Compounding a lack of ocean literacy is an overall lack scientific literacy. In the U.S., the level of high school graduate scientific literacy is well below the international average.

marine technology SOCIETY
Opportunity runs deep

Introduction

- Next Generation Science Standards (NGSS) in the United States offer an opportunity to incorporate technology and engineering in the K-12 curricula.
- At the collegiate level, hands-on experiences, internships and challenges are required to augment traditional lecture and laboratory experiences to better engage and prepare the future ocean workforce.
- The education of the 21st century ocean workforce requires not only a strong understanding of oceanography and other disciplines, but an ability to integrate science concepts, engineering methods and sociopolitical considerations.

marine technology
SOCIETY
Opportunity runs deep™

Science Education in the United States

- The U.S. continues to lose its competitive edge in math and science.
- STEM pipeline workforce issues.
- Graduation rates in technology-based studies remain low, and many fields do not have a diverse workforce.

marine technology
SOCIETY
Opportunity runs deep™

Opportunities for Improvement

- Students need to learn practically rather than theoretically, through self-motivated discovery as opposed to didactic lecturing and collaboratively using technology-based problem solving.

  "Children learn best if they are immersed in complex experiences and are given the opportunity to actively process what they have learned." -R.N. Caine

- Five common features of successful education programs that engage and prepare underrepresented students for technical careers: 1) *Defined outcomes, 2) Sustained commitment, 3) Personalization, 4) Challenging content and 5) Engaged adults.*

marine technology SOCIETY
Opportunity runs deep

MTS Contribution

Scholarships

- To date, MTS has awarded over $550,000 in scholarships to MTS Student members all over the world.

- The data collected suggests that the MTS scholarship program is of value to students, it is easy to apply for the scholarships, and the scholarship amounts are agreeable to the target population.

https://www.mtsociety.org/education/scholarships.aspx

marine technology SOCIETY
Opportunity runs deep

MTS Contribution

OCEANS Student Poster Competition

- The OCEANS Student Poster Competition, offers an opportunity for students to travel to an MTS/IEEE OES OCEANS conference to present their research, expenses paid.

- The Office of Naval Research (ONR) and the Office of Naval Research Global (ONRG) have supported the Student Poster Competition since inception.

marine technology
SOCIETY
Opportunity runs deep

MTS Contribution

Marine Technology Camps

Northwestern Michigan College – May 20 – 25, 2018

Activities include: ROVs, Sonar, Hydrographic Surveying, Research Vessel Operations

Registration: https://www.nmc.edu/resources/water-studies/

marine technology
SOCIETY
Opportunity runs deep

MTS Contribution

Marine Technology Camps

Rutgers University – June 3 – 8, 2018
Activities include: Glider operation, maintenance and piloting

Registration: https://rucool.marine.rutgers.edu/marine-technology-glider-camp

marine technology SOCIETY
Opportunity runs deep

MTS Contribution

Internship Program

- In 2018, MTS plans to begin a national (U.S. focused at first) MTS Internship Program for Undergraduate and Graduate level MTS student members.

- MTS will conduct preliminary applicant screening and provide the most qualified candidate applications to the participating member companies.

marine technology SOCIETY
Opportunity runs deep

MTS Contribution

Implement a Young Professionals Membership

MTS plans to create a new category of membership, a Young Professionals Membership to continue to engage early career professionals as they transition from student to professional status.

marine technology SOCIETY
Opportunity runs deep

MTS Contribution

Mentorship program

- MTS has launched a Marine Industry Mentoring Program (MIMP) for early career professionals.

- The MIMP is administered by means of an online match-making platform, accessible through the MTS web page

marine technology SOCIETY
Opportunity runs deep

MTS Contribution

K-12 Education and Outreach

- To reach K-12 students, MTS will develop a web site, Blue Futures. MTS's Blue Futures initiative will provide resources for students interested in marine technologies.

- A main component of Blue Futures will feature an education and career planning tool to help and encourage students in grades 7-12 to explore careers and understand the educational requirements to work in marine technology-related fields.

bluefutures

marine technology SOCIETY

Opportunity runs deep

MTS Contribution

Webinars

The MTS is in the process of offering webinars to the membership.

The first webinar series is a collaboration with the Gulf of Mexico Research Initiative (GoMRI) in support of the TechSurge (Monday, February 5).

The webinar series is focused on technological developments focused on improving response to future oil spills.

MTS techsurge

GULF OF MEXICO RESEARCH INITIATIVE

marine technology SOCIETY

Opportunity runs deep

bluefutures
marine technology
SOCIETY
Opportunity runs deep™

INVENTION EDUCATION:
FROM MATE ROV TEAM TO PATENT APPLICATION

SCIENCE
TECHNOLOGY
ENGINEERING
MATH

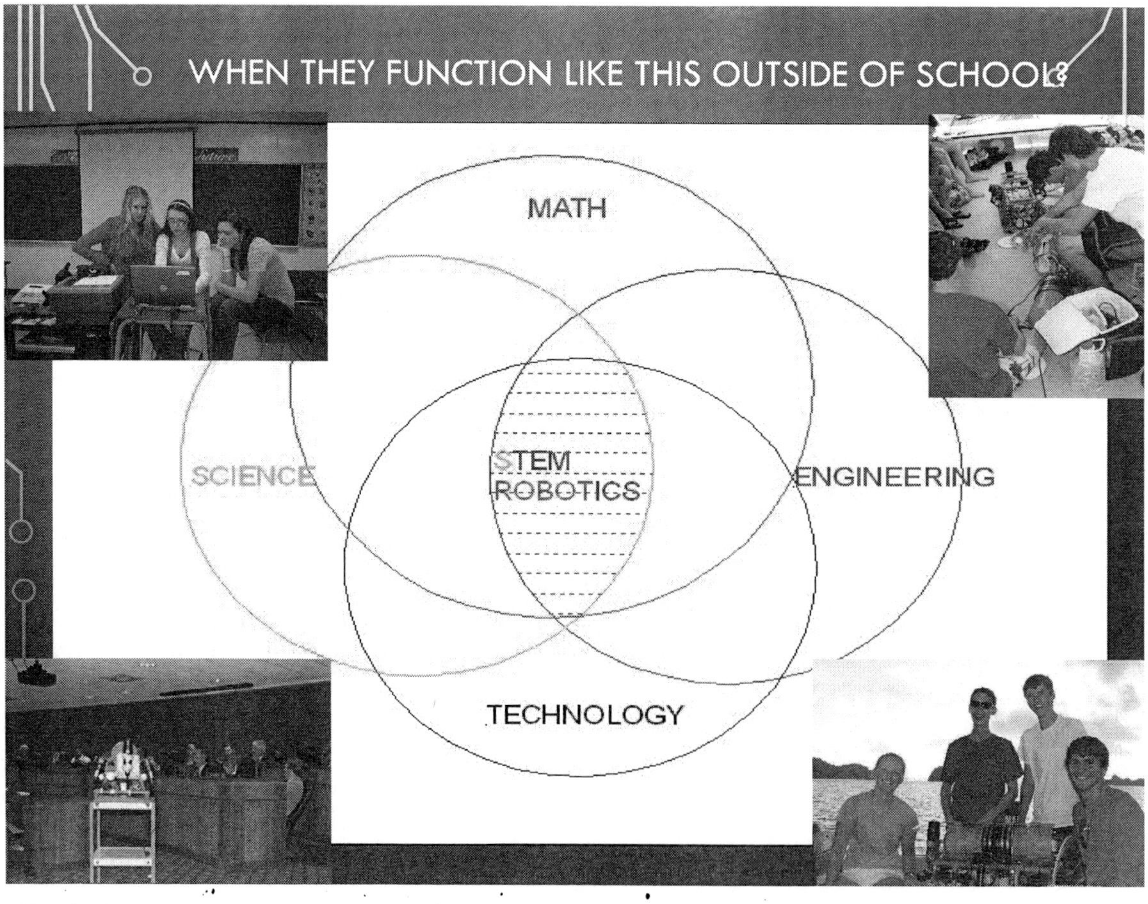
WHEN THEY FUNCTION LIKE THIS OUTSIDE OF SCHOOL?
MATH
SCIENCE
STEM
ROBOTICS
ENGINEERING
TECHNOLOGY

ROV COMPETITIONS

UNIVERSITY OF HAWAI'I AT

Neutral Buoyancy Laboratory
Operations Control Center
NBL
Visitor's Gallery

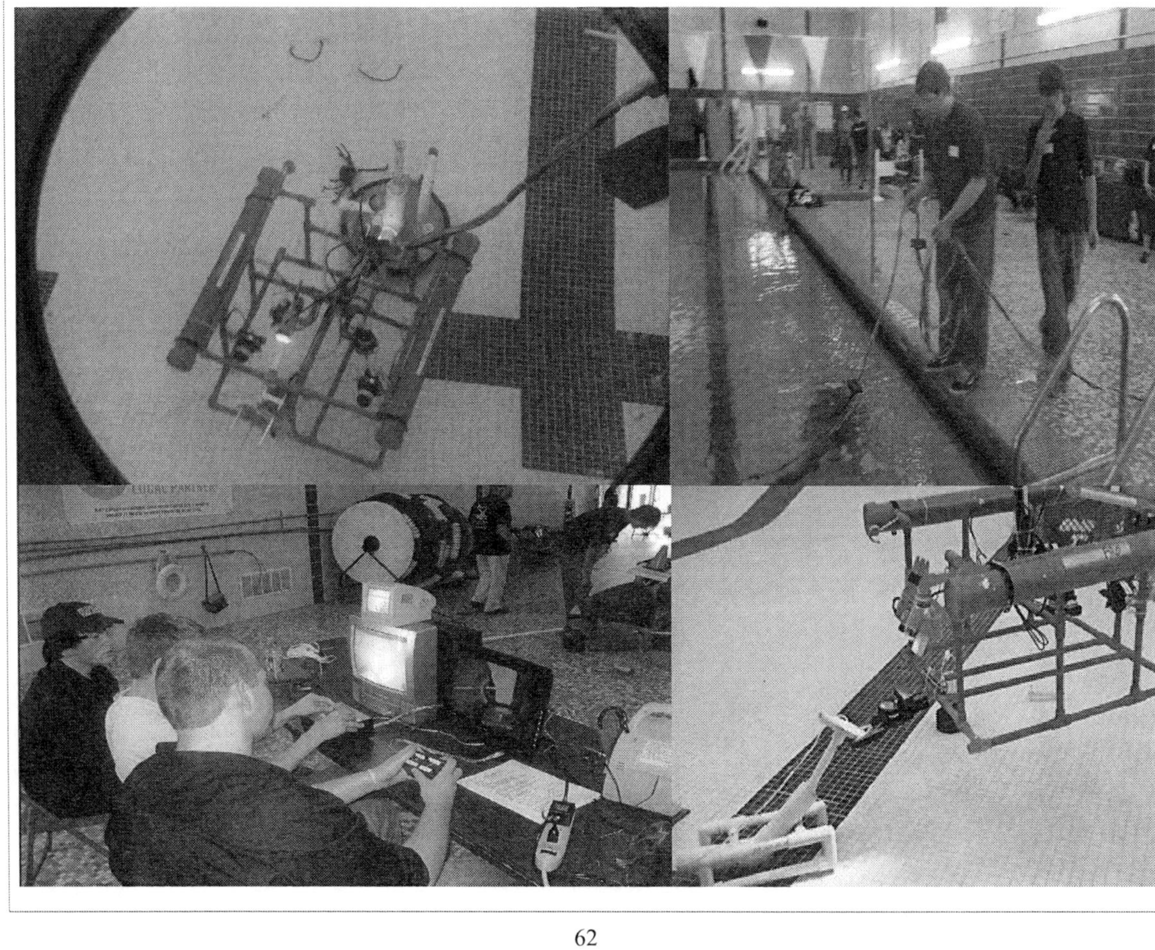

BENT PROP PROJECT
P-MAN EST. MCMXCIII

POW*MIA
YOU ARE NOT FORGOTTEN

AT THE GOING DOWN OF THE SUN AND IN THE MORNING, WE WILL REMEMBER THEM

BENT PROP PROJECT
P-MAN XIV
15 MARCH 2012 - 15 APRIL 2012
DINGAR
ROV
BENT PROP
PANTHER IV
STOCKBRIDGE HIGH SCHOOL
UNDERWATER ROBOTICS TEAM
PALAU EXPEDITION
22 MARCH 2012 - 07 APRIL 2012

https://www.youtube.com/watch?v=10rAjQT7uMM

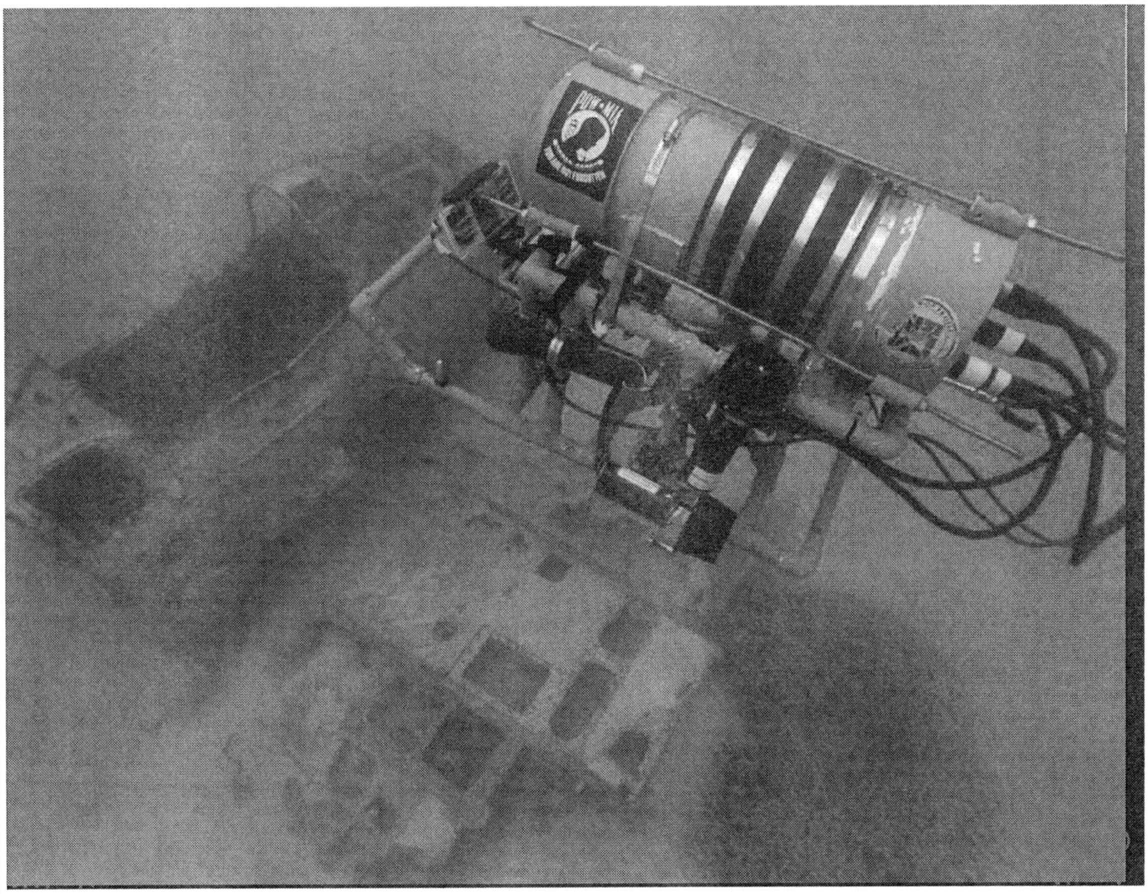

WORKING IN THE GREAT LAKES

GLERL

Great Lakes – Xmas Tree

Stockbridge Academic Excellence
LEMELSON-MIT
InvenTeams
Stockbridge High School
Stockbridge, Michigan
School of Engineering

A CHALLENGE FROM THE PRESIDENT

In 2013 President Remengesau challenged the team to find a way to use their technology to study and preserve the ocean, and coral reefs around Palau.

OUR FIRST PROTOTYPE

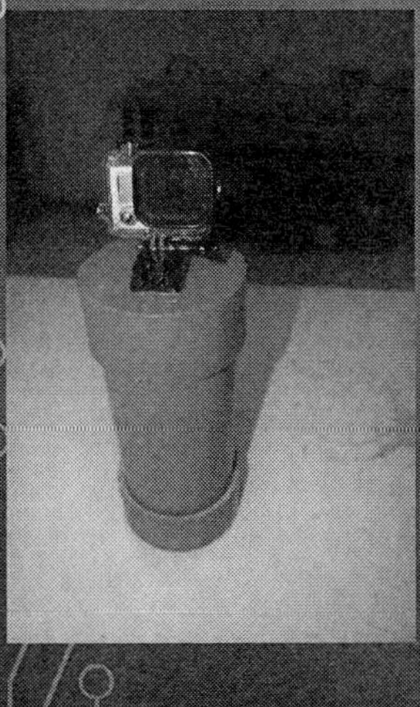

A Go Pro Camera attached to an external battery.

LEMELSON-MIT GRANT

- Lemelson-MIT InvenTeams™ are comprised of high school students, educators, and mentors that receive up to $10,000 each to invent technological solutions to real-world problems of their own choosing.
- The Stockbridge InvenTeam was awarded a Lemelson-MIT grant on October 14th, 2015.
- With this grant we are able to accomplish everything we need to make a successful product besides travel fare.

THE INVENTION

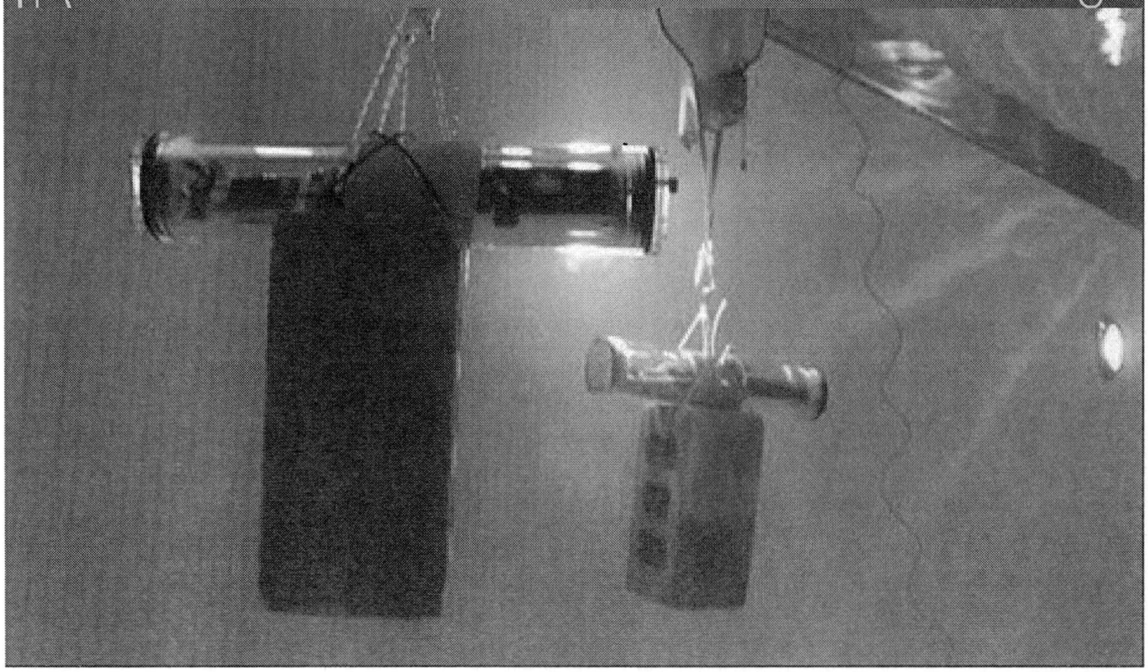

MIT

773

#MAKE WHAT'S NEXT

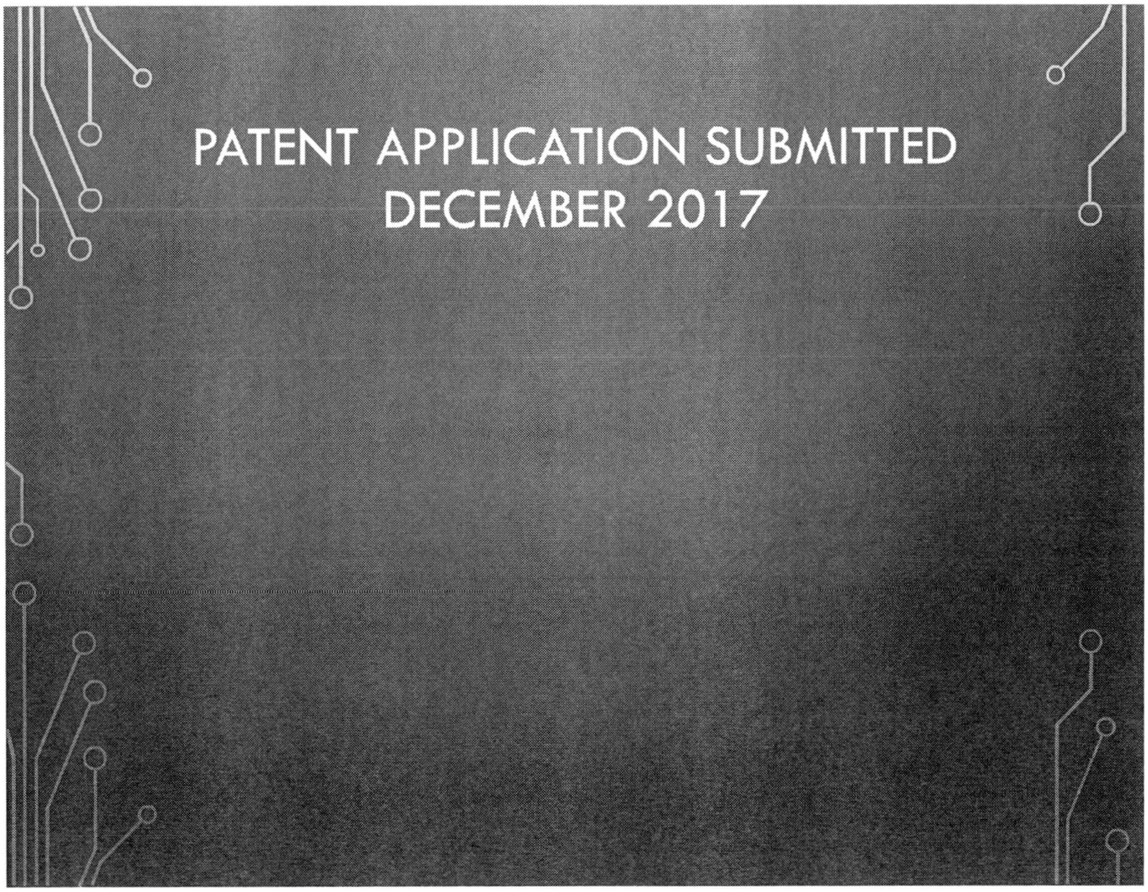
PATENT APPLICATION SUBMITTED
DECEMBER 2017

SPIN OFF CLASSES

- CAD & 3D Printing
- Digital Video Lab/Film Team
- JvInvenTeams/7th and 8th Grade STEM

3D PRINTING IN THE CLASSROOM

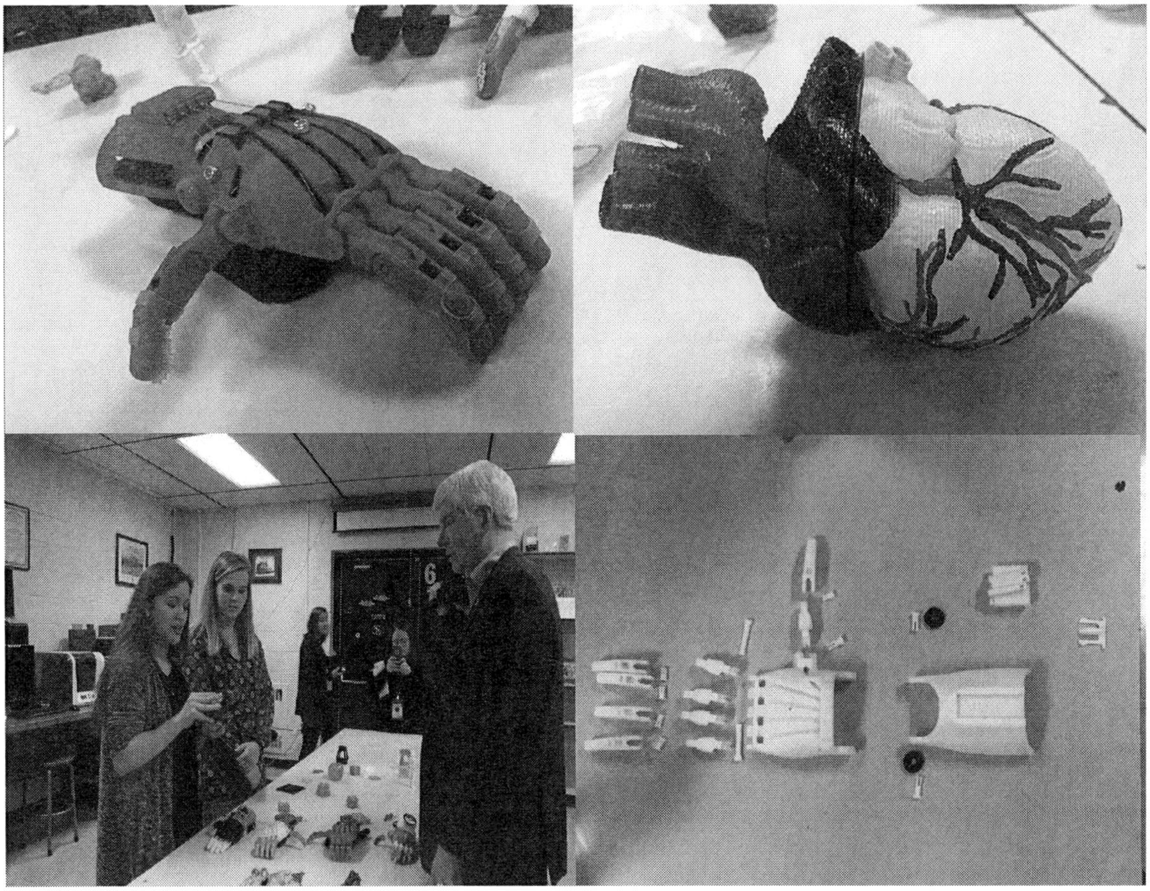

STOCKBRIDGE
FILMS

7TH & 8TH GRADE STEM
LEMELSON·MIT
JVInvenTeams
SHOE SOLES
SUPER LENS
CHILL OUT
NOISE MAKERS
GROWING GREEN
PUMP IT UP!
U CONTROL

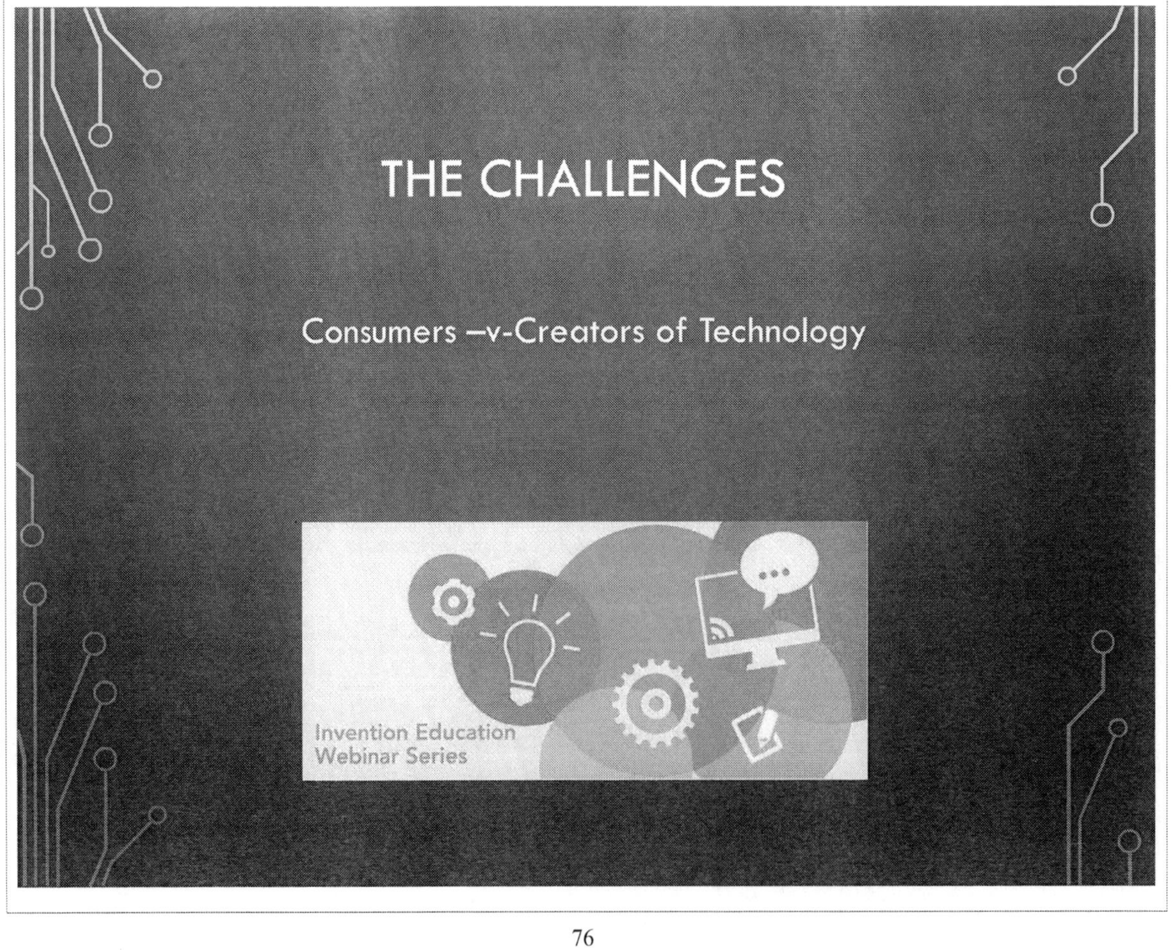
THE CHALLENGES
Consumers –v-Creators of Technology
Invention Education
Webinar Series

WHAT'S NEXT

- Fish ID software
- Swift-water rescue ROV
- Baited fish camera
- Palolo Worm Study

POTENTIAL RESEARCH QUESTION

- How do we get more girls interested in STEM?
- How is Stockbridge able to get more girls interested in STEM?
- How to we develop more female patent holders/applicants?
- Can we replicate our success?

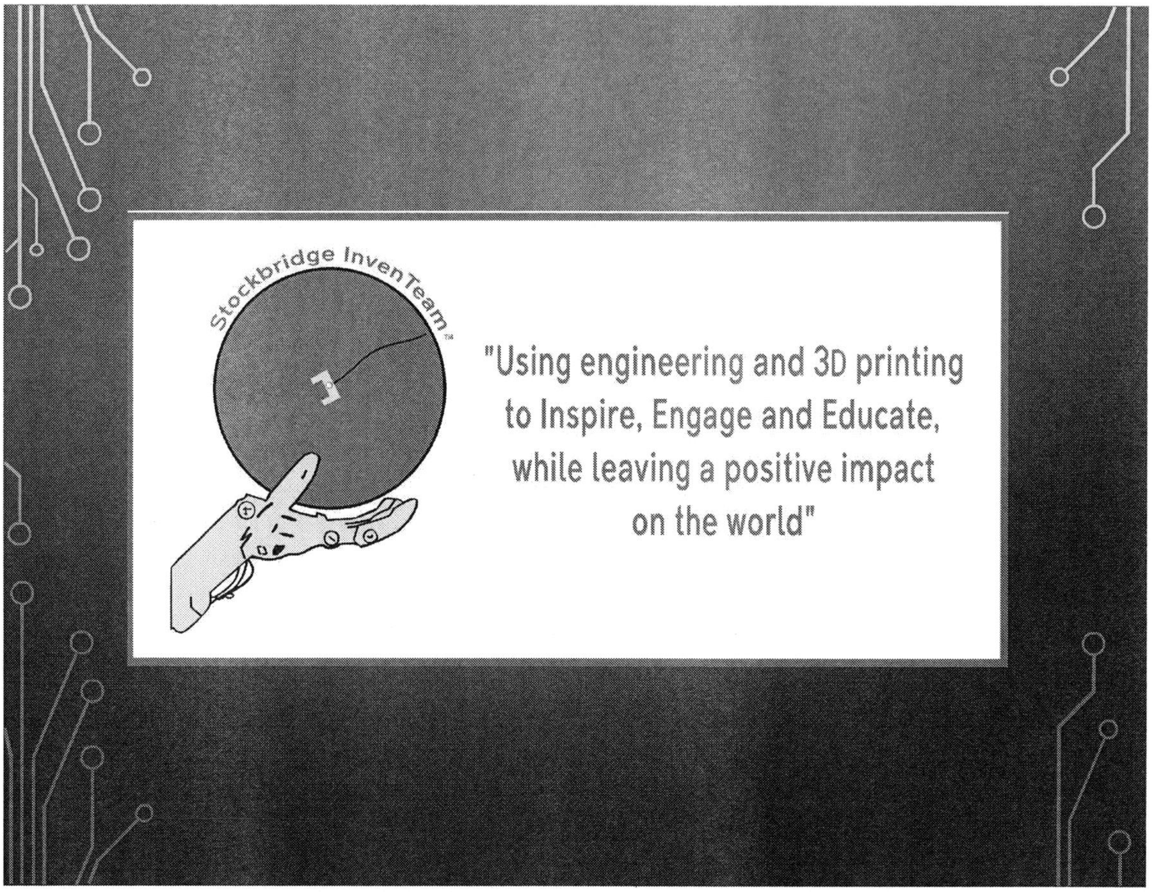

Stockbridge InvenTeam
TM
"Using engineering and 3D printing to Inspire, Engage and Educate, while leaving a positive impact on the world"

Stockbridge Robotics
Page
Notifications
Insights
Publishing Tools
Settings
Help
Stockbridge Robotics
@StockbridgeRobotics
Home
About
Photos
Reviews
Likes
+ Add a Button
Non-Governmental Organization (NGO) in Stockbridge, Michigan
Page Tips
How to Create Effective Posts
Help People Take Action
See All Page Tips
2,451 follows
Stockbridge Robotics
Tweets
Tweets & replies
Media
Introducing this years Stockbridge Robotics Marketing team. Team members are Kelly, Jake, Jillian, Katelyn, and Lauren!
Who to follow
Stockbridge Football
Trends

Divers Augmented Vision Display (DAVD)

UI2018
UNDERWATER INTERVENTION

NAVSEA
NAVAL SEA SYSTEMS COMMAND

07 February 2018

Authors: *Dennis G. Gallagher, Richard J. Manley, Allie M. Pilcher, William W. Hughes III*
Naval Surface Warfare Center-Panama City Division
Panama City, Florida USA

DISTRIBUTION STATEMENT A: Approved for public release; distribution is unlimited

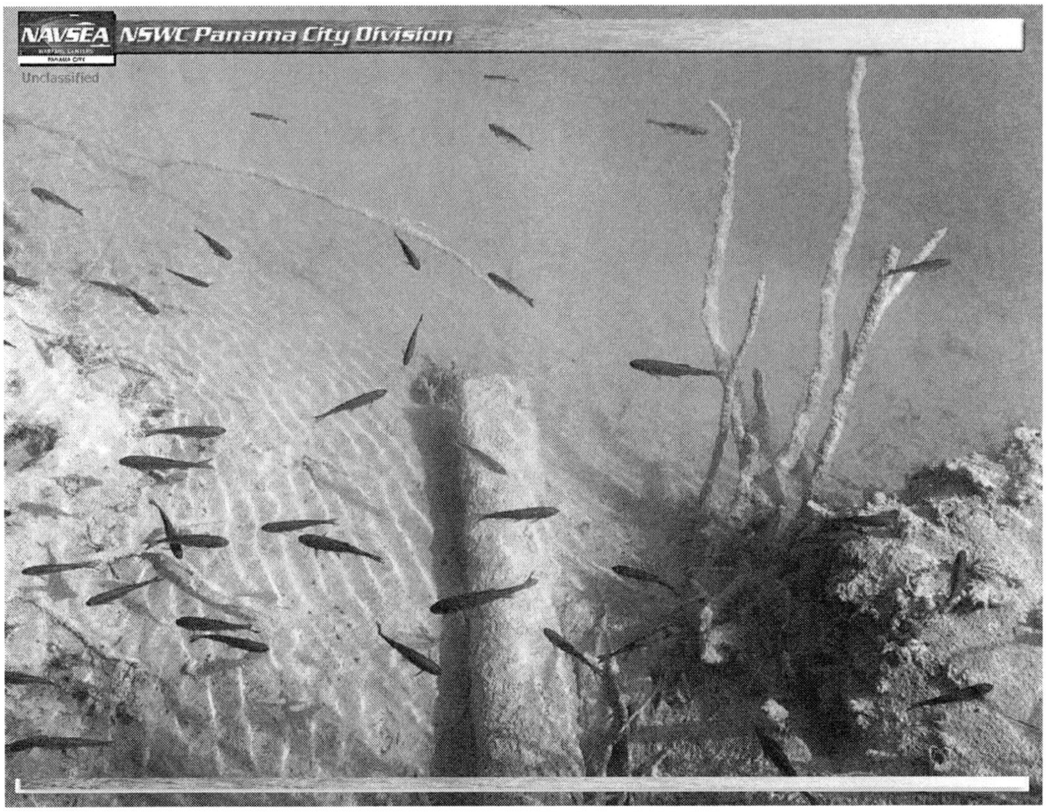
NAVSEA
NSWC Panama City Division
Unclassified

NAVSEA
NSWC Panama City Division
Unclassified
...except this is the real world
Unclassified

A Little History...

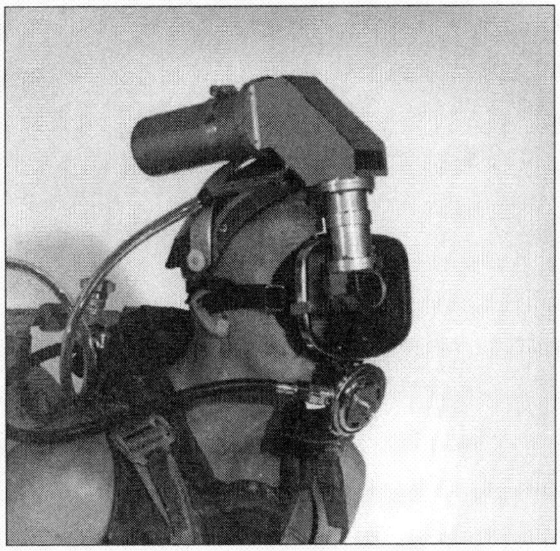

...circa 1960

Near-To-Eye (NTE) Diver Display Systems

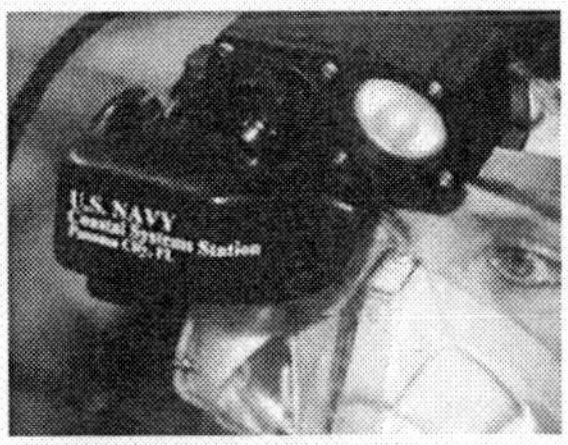

Video Night-Vision

Dive Computer Display

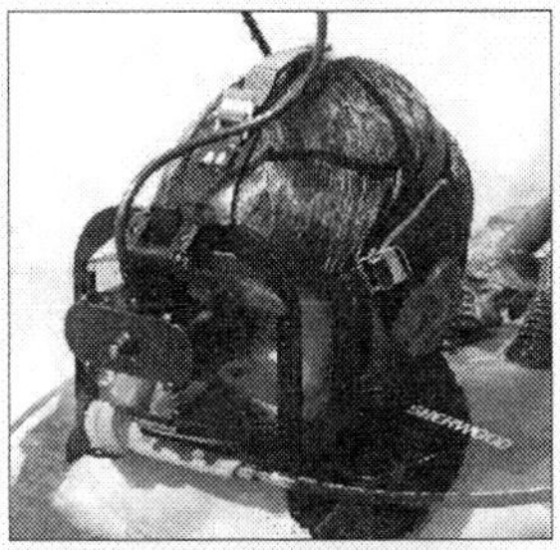

Binocular Sonar Display

Combat Diver NAV Module

How do we go from this...

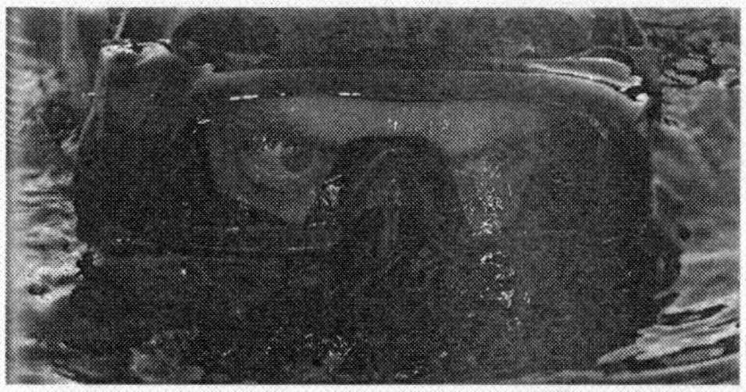

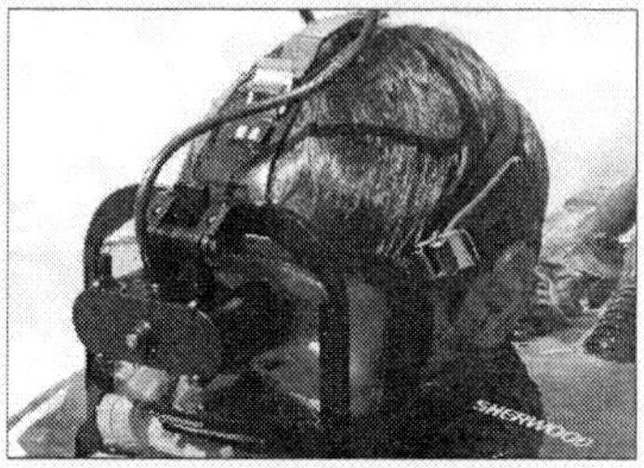

...to this

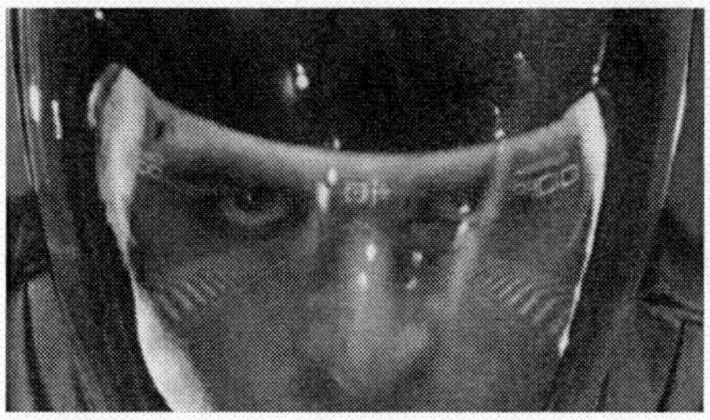

Military Aviation Head-Up Displays

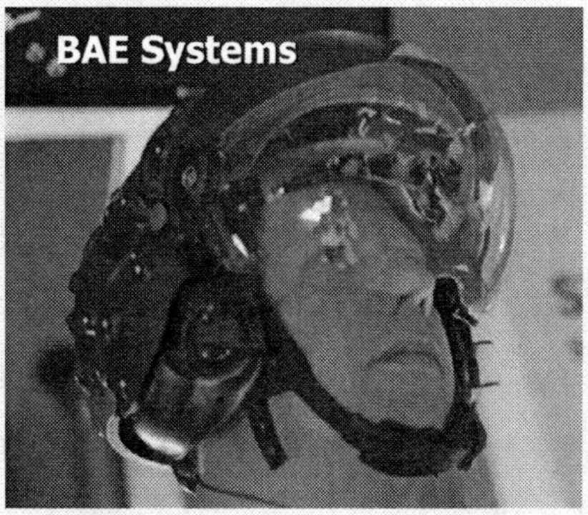

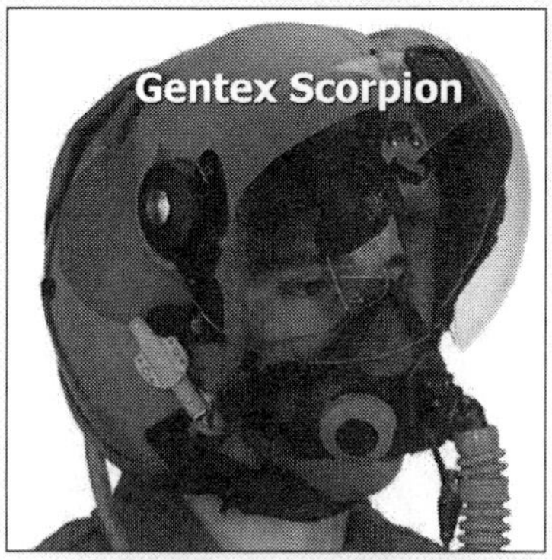

- Projection-Reflection Technology
- Higher Power requirement
- Larger size required

- Waveguide Technology
- Lower Power requirement
- Smaller modular configuration

Waveguide Optical Displays – The Basics

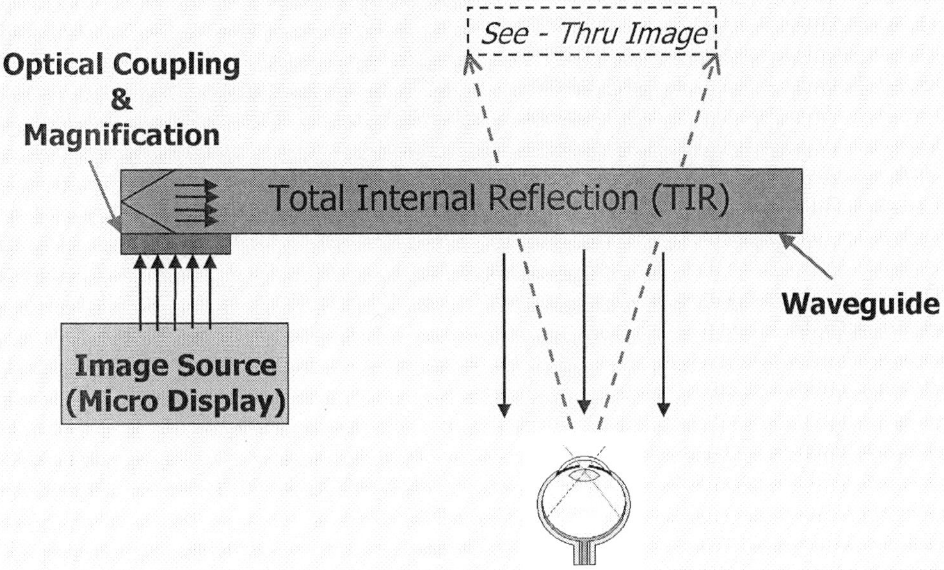

Lumus Optical (waveguide displays)

http://www.lumus-optical.com

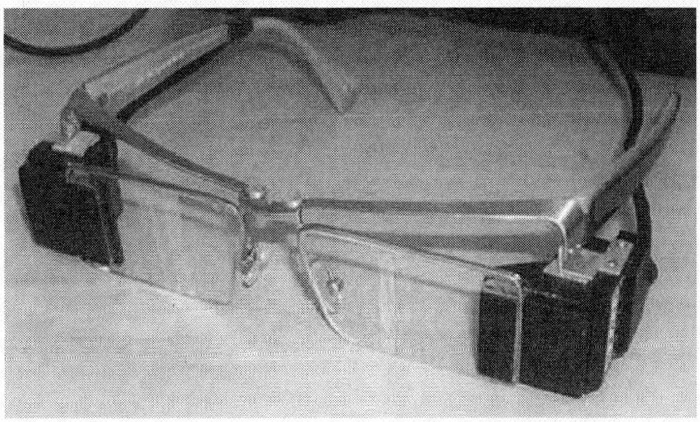

Laboratory Testing and Measurements

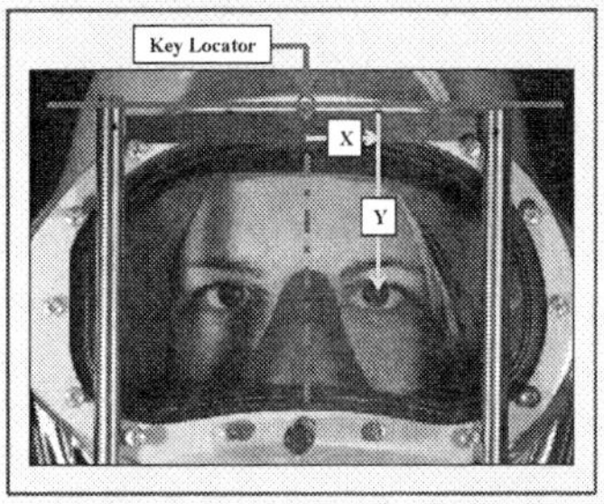

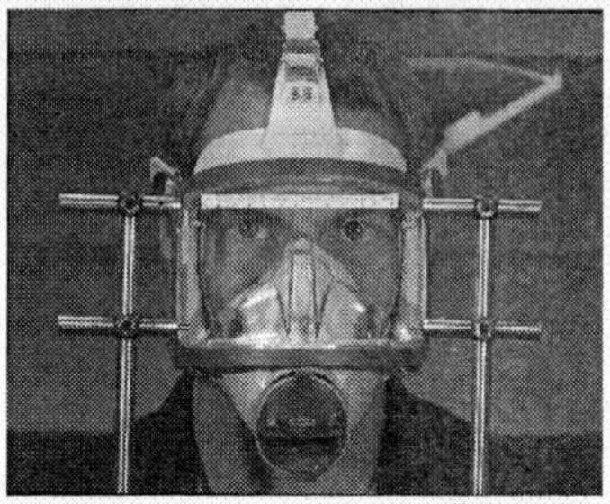

Eye Location measurements

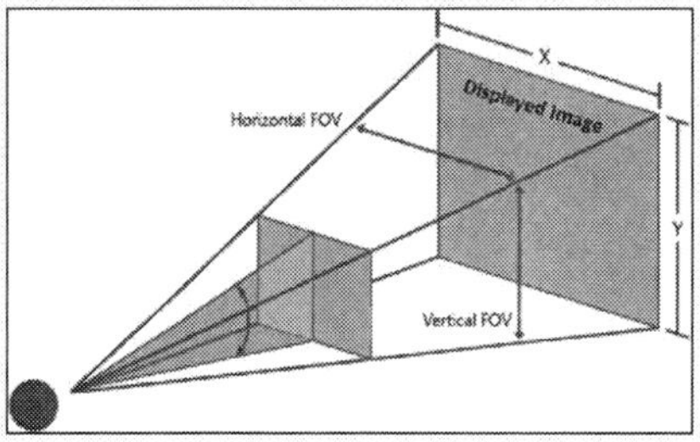

Focal Distance - Field-of-View – Eye Box – Image Convergence Distance

Design Reference Mission Simulations

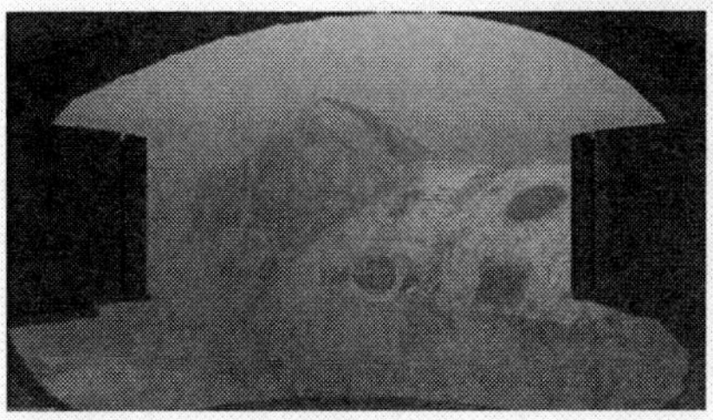

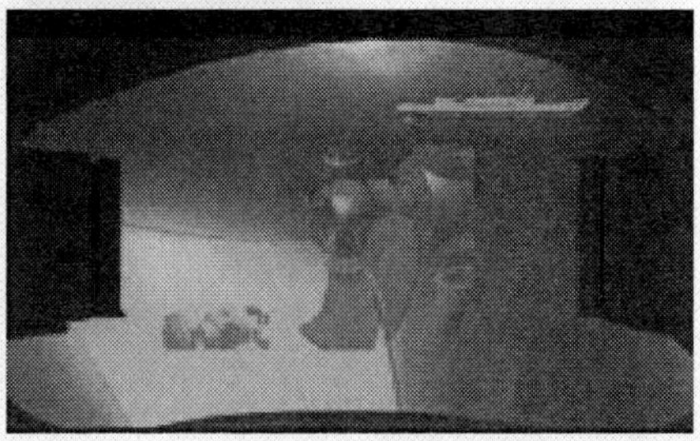

NAVSEA NSWC Panama City Division

Unclassified

Diver Evaluations

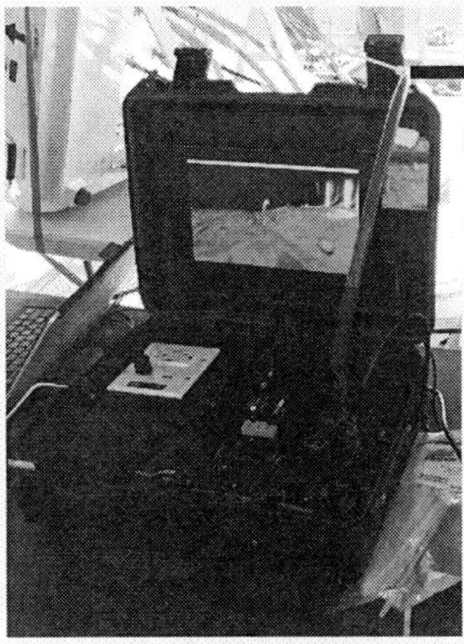

Topside Control Box

300 FT Umbilical

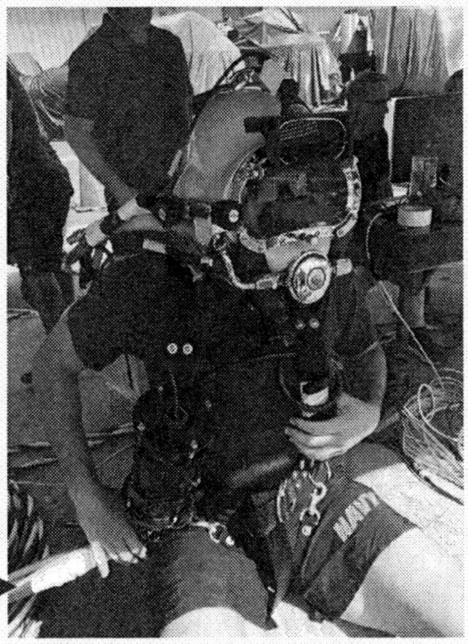

Unclassified

13

NAVSEA NSWC Panama City Division

Unclassified

Diver Evaluations (cont'd)

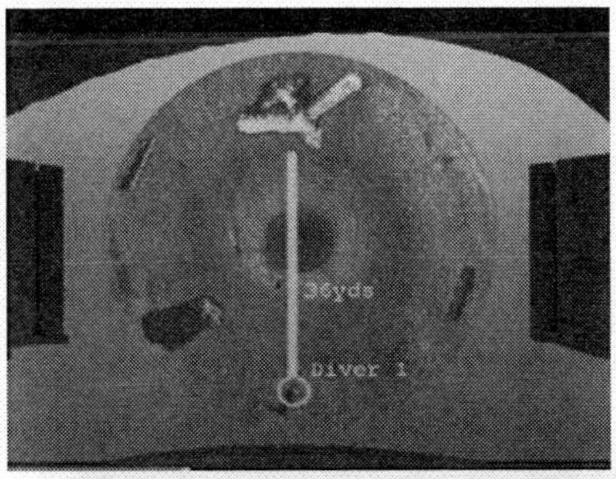

Navigation

Instruction

Photos

AR Overlays

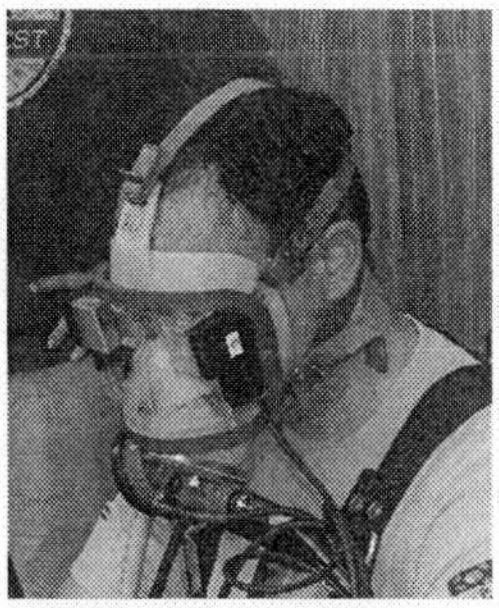

Unclassified

14

UNDERWATER INTERVENTION 2018

Questions?

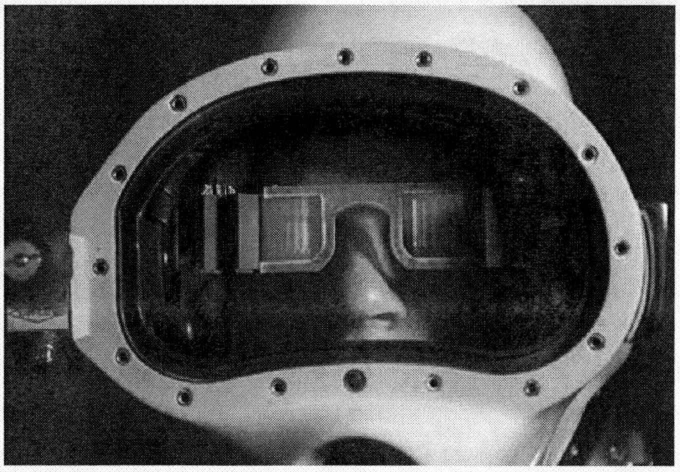

CONTACT:

Dennis G. Gallagher – Project Manager (Group S3D)
Naval Surface Warfare Center-Panama City Division
Underwater Systems Development & Acquisition Branch (Code E15)
Panama City, Florida 32407 USA
Tel: 850.235.5417/dennis.g.gallagher@navy.mil

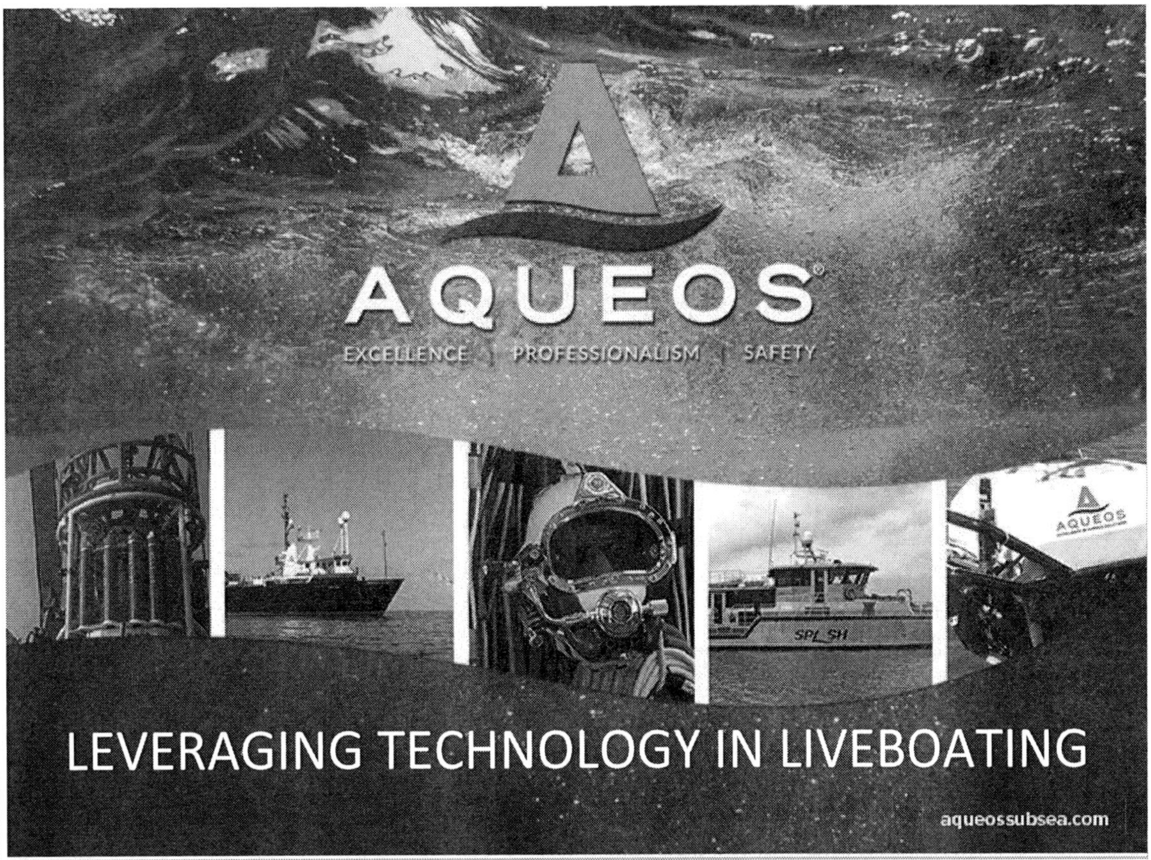

AQUEOS

OBJECTIVES:

- ➢Define Liveboating
- ➢Regulatory requirements/restraints
- ➢Industry Guidelines
- ➢Industry Best practices
- ➢Historical context of liveboating
- ➢Technology Advances
- ➢Enhanced capabilities of modern spread

EXCELLENCE | PROFESSIONALISM | SAFETY aqueossubsea.com

AQUEOS

LIVEBOATING DEFINED:

- USCG & OSHA define liveboating as supporting a surface supplied (air) diver (or mixed gas diver) from a vessel (which is) underway
- ADCI defines liveboating as a diving technique where a single surface supplied diver performs work underwater while his hose is tended from the bow of a vessel which is manually operated by the vessel master and underway using its main propulsion system.
- IMCA defines it as the practice of supporting a diver for a non-DP vessel which is under power and making way

EXCELLENCE | PROFESSIONALISM | SAFETY   aqueossubsea.com

AQUEOS

LIVEBOATING vs. DP DIVING:

- USCG is completely silent on the matter of DP diving, although in accordance with its definition, USCG has required a variance or waiver to permit DP diving during nighttime operations – it is widely assumed this will be addressed in new regulations in our lifetime - maybe
- ADCI and IMCA are clear in their intent by their definitions that DP diving is not liveboating.
- DP operations are widely considered within our industry as a form of "mooring" for a vessel. With the vessels redundant reference systems and other safeguards, a DP vessel is considered as safe as a 4-point vessel on anchors or a vessel alongside a structure

EXCELLENCE | PROFESSIONALISM | SAFETY   aqueossubsea.com

AQUEOS

REGULATORY REQUIREMENTS:

- USCG proscribes liveboat diving under the following conditions:
  - in seas impacting the vessel's ability to hold station or the diver's mobility or work function
  - During periods of restricted visibility and at night
  - In depths in excess of 220 fsw or when in-water decompression exceeds 120 minutes

EXCELLENCE | PROFESSIONALISM | SAFETY

aqueossubsea.com

AQUEOS

REGULATORY REQUIREMENTS (cont):

- USCG requires the following for liveboating:
  - A standby diver be available while the diver is in the water
  - The diver must have a reserve breathing gas supply
  - A boat is ready to be launched in the event of an emergency
  - The vessel propellers must be stopped for diver entry and exit
  - The person in charge is to be notified prior to diver entry and exit of the water
  - A MEANS IS USED TO PREVENT THE DIVER'S HOSE FROM ENTANGLING IN THE PROPELLERS OF THE VESSEL

EXCELLENCE | PROFESSIONALISM | SAFETY

aqueossubsea.com

EOS

AQUEOS

INDUSTRY GUIDELINES:

- ➢IMCA simply states "Liveboating", which is the practice of supporting a diver from a non-DP vessel which is under power and making way, SHOULD not be used.
- ➢IOGP agrees with IMCA. An effort on the part of a couple operators to have language on liveboating accepted died on the vine.

AQUEOS

INDUSTRY GUIDELINES (cont.):

- ADCI Rev 6.2 allows Liveboating with the following provisions:
  - USCG regulations apply except as superseded below
  - Maximum depth allowed is 130 fsw
  - NO planned in-water decompression allowed
  - Only one surface powered tool can be deployed at a time
  - A "kill" switch is required in immediate vicinity to vessel operator to instantaneously shutdown the engines
  - A mechanical devise is required to prevent umbilical entanglement in the vessel propulsion

AQUEOS

INDUSTRY BEST PRACTICES:

- The US Gulf of Mexico Diving Safety Work Group (DSWG) :
  - Drafted a Report in July of 2013 on Liveboating which included:
    - A gap analysis between ADCI, IMCA, USCG and OSHA
    - A listing of industry known/reported incidents and near misses relative to liveboating with underlying root causes and corrective actions
    - A comprehensive hazard analysis with potential negative outcomes and mitigations
    - A listing of recommended emergency drills to be conducted

AQUEOS

INDUSTRY BEST PRACTICES (cont.):

➢The US Gulf of Mexico Diving Safety Work Group (DSWG) :

- ➢ Recommends evaluation of other options for site and work specific diving mode
- ➢ Places heavy emphasis on having auditable equipment and vessel maintenance programs in place
- ➢ Places heavy emphasis on master and supervisor experience in liveboating operations
- ➢ Requires a detailed HAZID be conducted considering site specific weather related conditions, crew experience, depth, decompression obligations, vessel traffic, work scope, SIMOPS, etc.
- ➢ Recommends the use of diver tracking and shaft revolution counters

EXCELLENCE | PROFESSIONALISM | SAFETY

aqueossubsea.com

AQUEOS

HISTORICAL CONTEXT OF LIVEBOATING:

➢In its "Glory Years" Liveboating was primarily used to inspect newly laid pipelines prior to ROV's

- ➢ Typically used spot market utility boats adapted for the ... od ... s, e

EXCELLENCE | PROFESSIONALISM | SAFETY

aqueossubsea.com

AQUEOS

EXCELLENCE | PROFESSIONALISM | SAFETY

aqueossubsea.com

AQUEOS

HISTORICAL PERSPECTIVE:

- ➢In the past 30 years there have been ZERO serious documented liveboating incidents in the Gulf of Mexico or Alaska
- ➢On operator recently conducted an internal review of liveboating dives with approximately 1,100 dives per calendar year over a five year period without a single "liveboating" related incident
- ➢Compare those statistics to underwater burning, excavation and differential pressure – all of which are recognized as high risk but manageable with adequate controls

EXCELLENCE | PROFESSIONALISM | SAFETY

aqueossubsea.com

AQUEOS

WHY LIVEBOAT?

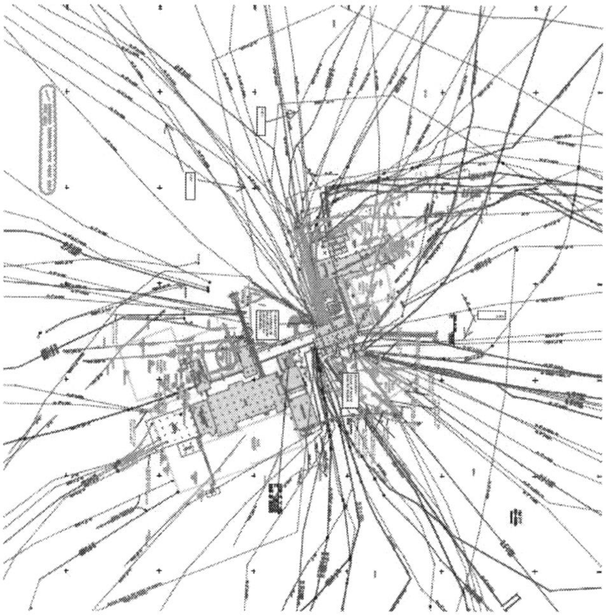

Field congestion prohibits anchoring or spudding or pads

Shallow depths are not conducive to DP operations

It is necessary in Alaska due to the rapid and 30'+ tidal changes.

EXCELLENCE | PROFESSIONALISM | SAFETY

aqueossubsea.com

TECHNOLOGY ADVANCES:

Use of water jet propulsion in lieu of propellers with intake guards to prevent umbilical or tooling hoses to enter intake

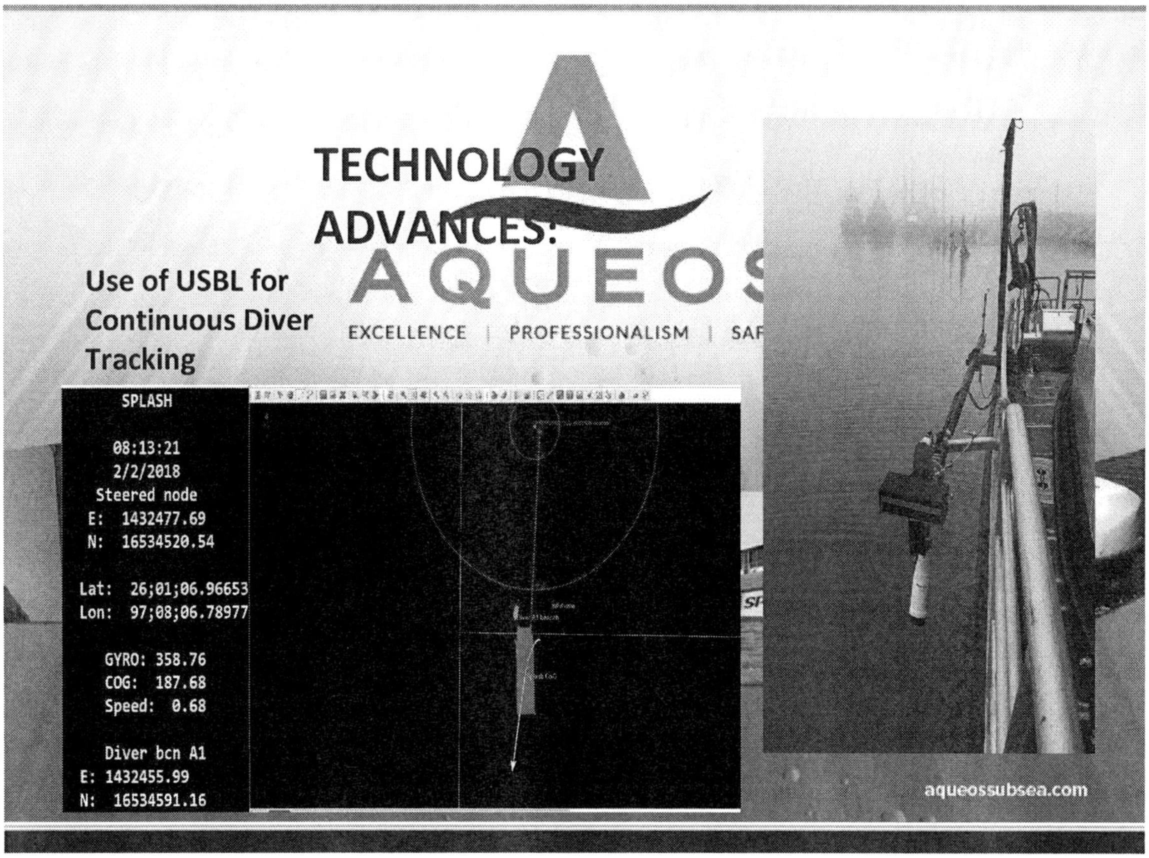
TECHNOLOGY ADVANCES:
Use of USBL for Continuous Diver Tracking
AQUEOS
EXCELLENCE | PROFESSIONALISM
SPLASH
08:13:21
2/2/2018
Steered node
E: 1432477.69
N: 16534520.54
Lat: 26;01;06.96653
Lon: 97;08;06.78977
GYRO: 358.76
COG: 187.68
Speed: 0.68
Diver bcn A1
E: 1432455.99
N: 16534591.16
aqueossubsea.com

TECHNOLOGY ADVANCES:
Vessel Design with Low Bow for Diver Entry
EXCELLENCE
Diver in the water
aqueossubsea.com

AQUEOS

TECHNOLOGY ADVANCES:

Use OF NITROX diving to minimize decompression obligations – safety to the diver

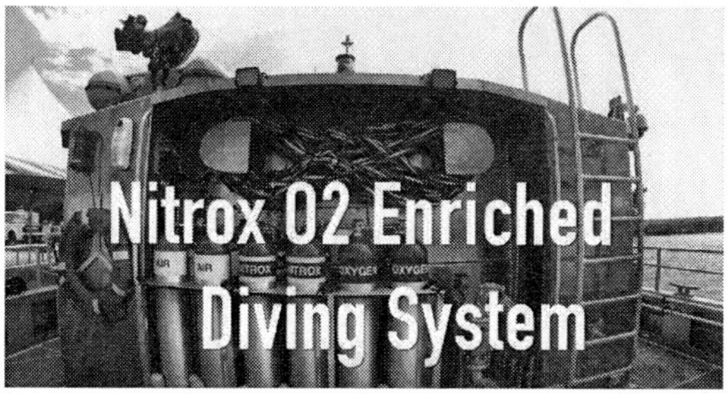

EXCELLENCE | PROFESSIONALISM | SAFETY

aqueossubsea.com

TECHNOLOGY ADVANCES:

AQUEOS

EXCELLENCE | PROFESSIONALISM | SAFETY

Climate Controlled Decompression chamber

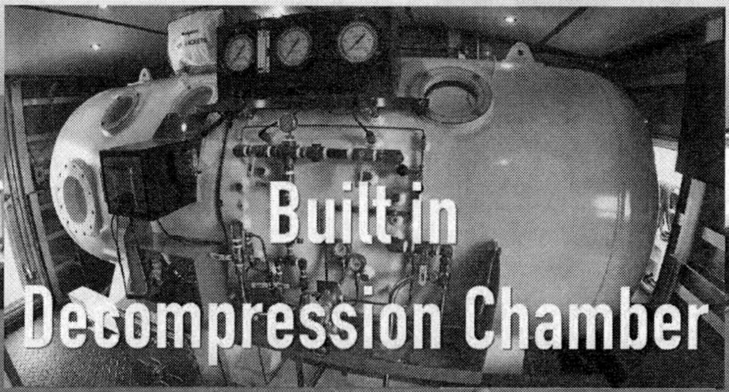

aqueossubsea.com

TECHNOLOGY
ADVANCES:
Enhanced Supervisor Dive Controls and Visibility
EXCELLENCE
360 Viewing from Pilot House
aqueossubsea.com

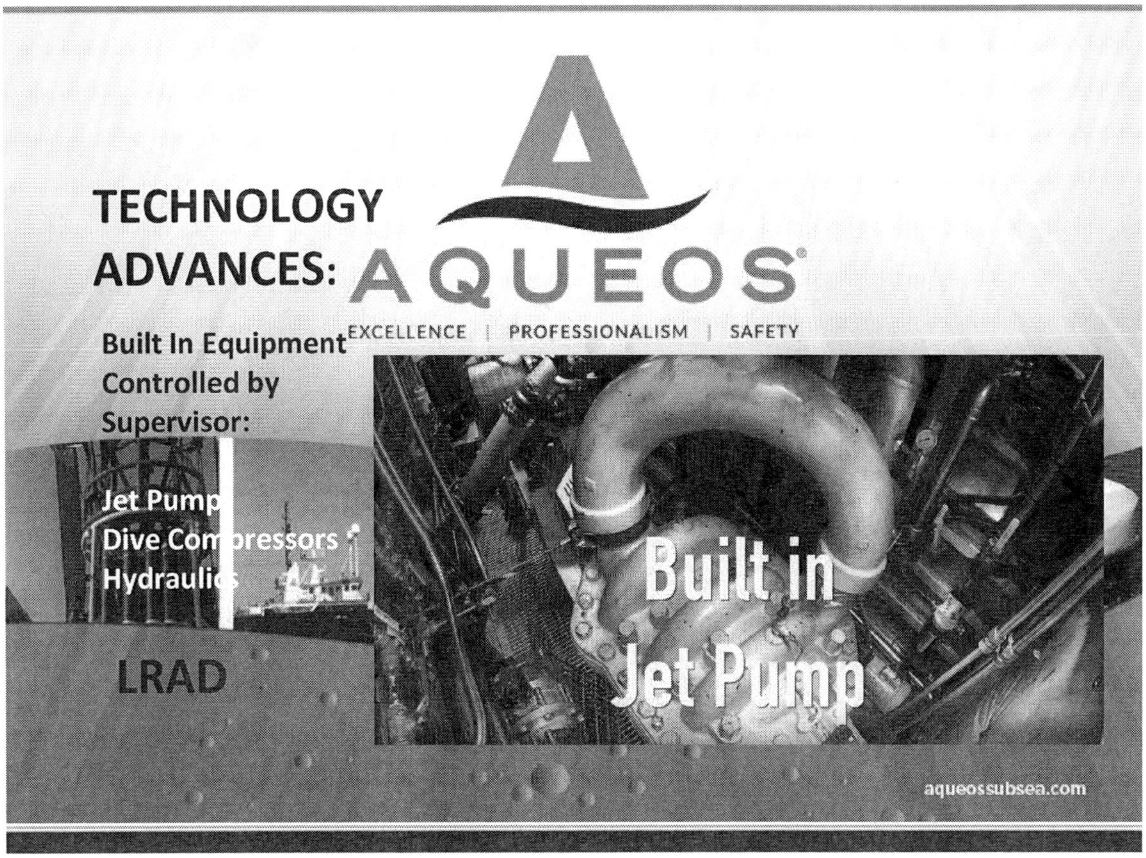
TECHNOLOGY
ADVANCES:
AQUEOS
EXCELLENCE | PROFESSIONALISM | SAFETY
Built In Equipment Controlled by Supervisor:
Jet Pump
Dive Compressors
Hydraulics
LRAD
Built in Jet Pump
aqueossubsea.com

OTHER ENHANC[redacted]TIES:
AQUEOS
EXCELLENCE | PROFESSIONALISM | SAFETY
Maneuverability
Speed (20k cruising)
Precision controls
Mooring 3-point
Active Mooring

[redacted]:
aqueossubsea.com

MATE Competition Update

Jill Zande
Associate Director & Competition Coordinator, MATE Center
President, MATE II

MATE

MATE II

AT-A-GLANCE

- Began in 2001, in partnership with the MTS ROV Committee
- Has involved well over 17,000 students in grades 4-16 who work in teams to tackle missions based on the ocean STEM workplace
- Includes one international competition and a network of 31 (and growing!) regional contests
- Involves 100s of working professionals & organizations
- Includes underwater missions and technical reports, engineering presentations, and marketing (poster) displays
- Gets students excited about STEM and motivates them to problem-solve, think critically, manage a project, and work together as a team
- Is encouraging students to pursue STEM degrees and jobs in the field
- Was featured in a book, documentary, and Hollywood movie

MATE

2017 marked the 16th international competition!

MTS
marine technology society

COMPETITION REACH

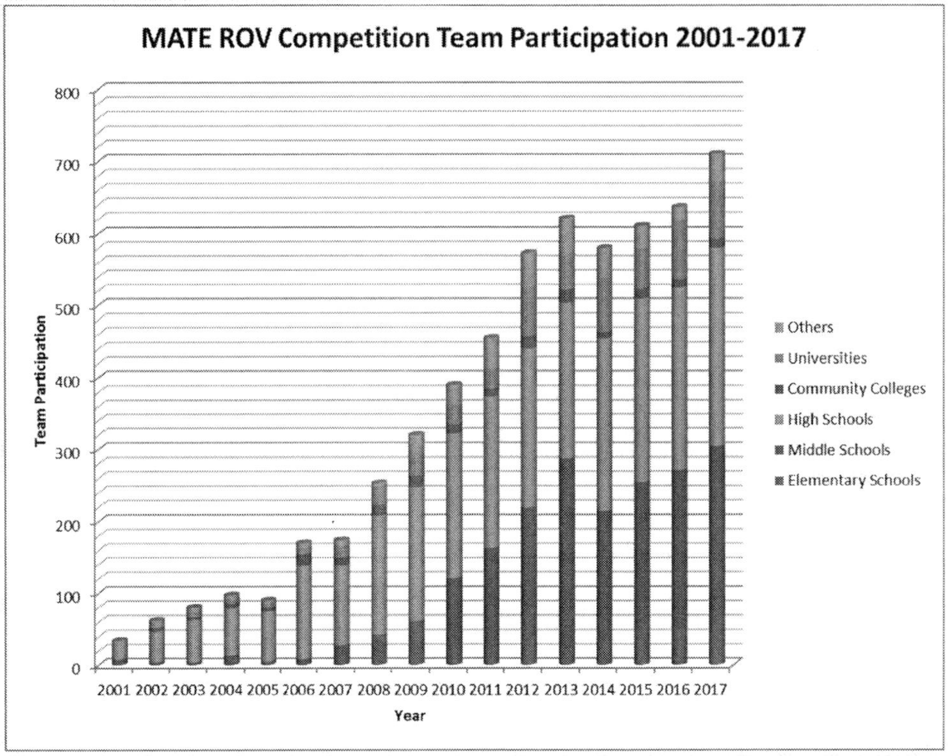

2017 NUMBERS

- 710 teams from 20 countries (including the U.S.) around the world
  - 258 SCOUT, 86 NAVIGATOR, 273 RANGER, 93 EXPLORER
- 4,300 students and 1,100 mentors
- Based on surveys completed, 19 regionals and 1 international had nearly 300 volunteers serve as judges, technical support, and more

STUDENT SURVEY SUMMARY RESULTS

STEM Careers

- After building their ROVs...(N=1,848)
  - 78% more aware
  - 77% more interested

STEM Knowledge/Skills

- Because of their ROV projects increased their skills & knowledge in...(N=1,848)
  - 92% engineering
  - 91% technology
  - 81% science
  - 62% math

21st Century Skills

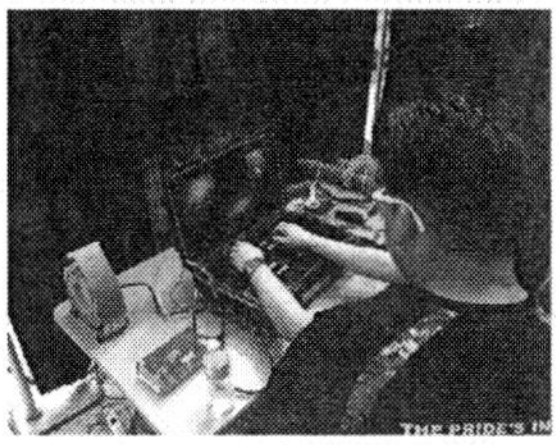

- Participating in the ROV project improved their...(N=1,848)
  - 84% problem solving
  - 84% teamwork
  - 80% critical thinking
  - 72% leadership skills
  - 66% organizational skills

STUDENT SURVEY SUMMARY RESULTS

Ability to communicate engineering design process

- 82% (N=1,848) stated that participating helped them learn how to communicate their engineering design to other people

Ability to apply STEM to real-world problems

- 86% (N=1,848) indicated that participating helped them learn to apply STEM to real world problems

Quotes:

I loved competing! I learned both technical skills and leadership skills, and it inspired me to pursue a career in engineering.

I think ROV has taught more than just academics, it also taught teamwork and coo peration in the business world.

The experience, knowledge, skills, and friendships that have been gained and developed through this program have proved invaluable and life changing.

TEACHER/MENTOR SURVEY SUMMARY RESULTS

STEM & 21st Century Skills

- 98% (N=382) observed their students had increased their knowledge and skills
- 96% (N=382) observed their students had increased their team building, problem solving, and/or critical thinking

PARENT SURVEY SUMMARY RESULTS

- Building an ROV contributed to improving their children's grades in...(N=537)
  - Engineering/robotics (53%)
  - Science (54%)
  - Math (45%)
  - Computers (45%)

JUDGE SURVEY SUMMARY RESULTS

- 95% (N=283) responded that the ROV competition provided a valuable venue to prepare their students for STEM careers

WHERE ARE THEY NOW?

2015 ALUMNI SURVEY (N=432)

- 85% of college degrees (N=220) were STEM degrees
- 85% of college students (N=236) were STEM majors
- 73% of employed alumni (N=320) had a STEM-related job
- 67% (N=432) credit the program with influencing their education/career
- The program played a role in attaining employment (37%), college admittance (36%), internships (30%), awards (21%), and scholarships (21%)

2017 DATA MATCH WITH THE NATIONAL STUDENT CLEARINGHOUSE

- 3,974 former competition participants were identified for matching
- 66% were found in the NSC database
  - Of the 1,779 with a major of study in the database, 66% were enrolled in a STEM major
  - 76% of the 828 degrees were awarded in STEM disciplines

WHERE ARE THEY NOW?

2017 ALUMNI PANEL

- Virgin Voyages
- General Motors
- Cisco Systems
- Sea-Bird Electronics
- Disney
- UW (2018 MTS ROV Committee Scholarship)

AT UNDERWATER INTERVENTION

- Blue Robotics (booth #640)
- SeaDrone (booth #233)
- OpenROV (scoping the exhibit floor)

WHERE ARE *WE* NOW?

- Planning to launch another alumni survey later this spring
- Submitting another set of data to the NSC
- Working to increase diversity and accessibility
- Looking at the impact of multi-year participation
  - Additional analyses to determine linkages between multi-year participation and pursuit of post-secondary education and choice of college major and STEM career.
- Looking at effective ways to manage growth

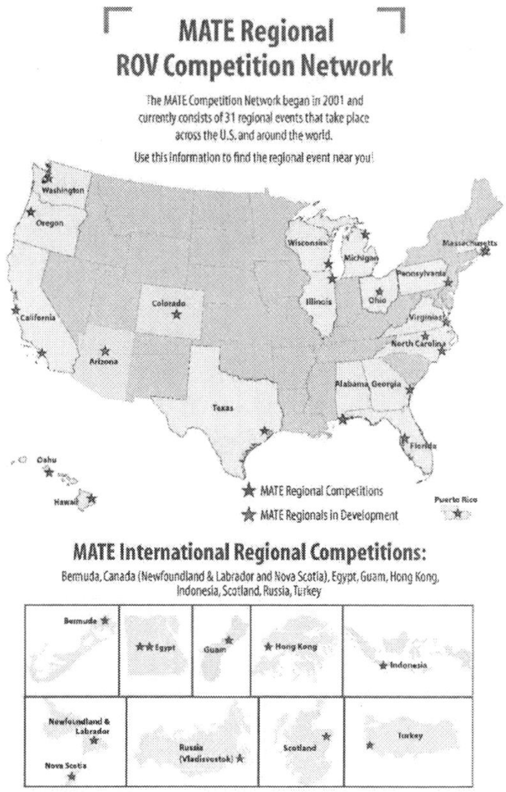

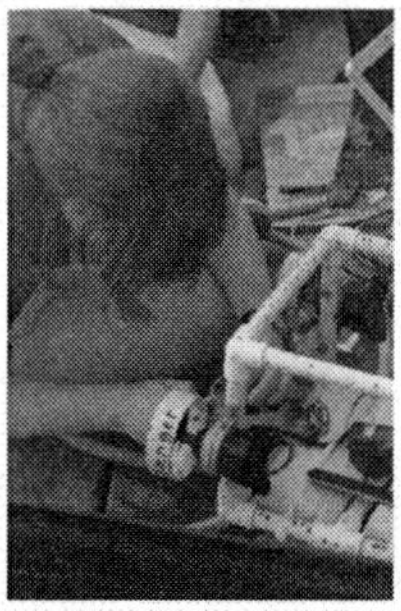

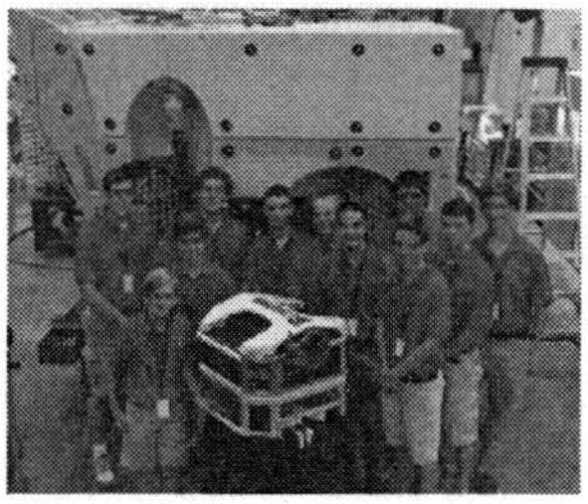

MATE Inspiration for Innovation (MATE II)

- IRS 501(c)3 status last April
- Opening doors
  - Boeing Company volunteer matching
  - GM Dollars for Donors
  - Motorola Solutions Foundation, etc.
  - Global Good Foundation/Intellectual Ventures
- Teams are benefitting as well
  - Certain states offer funding to schools that participate in robotics competitions organized by 501(c)3s

Questions?

Hope to see you at the 2018 international competition June 21 – 23 at the King County Aquatic Center in Federal Way, WA or at a regional event near you!

Jill Zande
MATE Center/MATE II
jzande@marinetech.org
(831) 646-3082
www.marinetech.org

Competition FB page: @materovcompetition
Twitter: @matecenter
#MATEROV

Unmanned Maritime Systems for Oceanographic Applications and Hydrographic Missions

Michael Bendzlowicz
USW Technical Lead N93
Naval Meteorology and Oceanography Command

This brief is UNCLASSIFIED
Approved for Public Release

Naval Oceanography — *Observe-Predict → Fight-Win*

America's Navy: Around the World, Around the Clock

Naval Oceanography — *Observe-Predict → Fight-Win*

Increasing Competition

Naval Oceanography — Observe-Predict → Fight-Win

Stronger, Faster, Smarter, Better

"Our adversaries are not going to hand victory to us – we're going to have to fight hard to win it."

Admiral John Richardson
Chief of Naval Operations

Naval Oceanography — Observe-Predict → Fight-Win

Asymmetric Advantage: Naval Oceanography

Naval Oceanography — *Observe-Predict → Fight-Win*

Naval Oceanography Strategy

Electromagnetic Maneuver Warfare

Advancing Support for Distributed Ops

Undersea Warfare

Improving to Beat the Competition

Unmanned Systems

Winning the Away Games

People

People First Mission Always

Positioning, Navigation and Timing

Assuring Distributed Operations

Resource Protection

Expanding for More Sensitive Platforms

Naval Oceanography — *Observe-Predict → Fight-Win*

Naval Oceanography Inventory

Autonomous Underwater Vehicles

REMUS 6000s - 4
REMUS 2500s - 2
REMUS 600s - 8 *
REMUS 100s - 3

-1,800 Missions -58,000 Nautical Miles
-18,000Hours (661 Days)

* 2 REMUS 600 currently under acceptance testing

Gliders

136-in inventory
35 - Max gliders deployed at one time

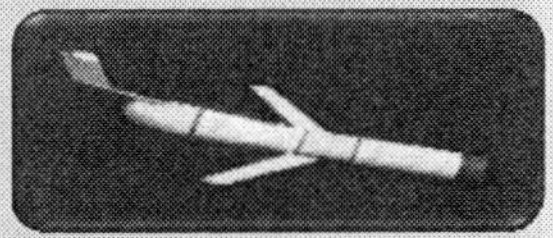

- 177,000 miles traveled by gliders
- 335 Operational Deployments

Warfare Specific Inventory

10 - MK18 Mod 1/ REMUS 100 UUVs
2 - SEABOTIX ROVs
22 - Ivers (20 NOSWC/2 FST)
UAVs (Pumas, ScanEagles and quadcopters)
- UUVs traveled 972.87nm
- 147.38 nm^2 of ocean bottom surveyed

Others

Surface Drifting Buoys - 74
Profiling Floats – 183
Wave Gliders - 5

20 Years Over 200,000 Miles 20 Different Platforms

Naval Oceanography Observe-Predict → Fight-Win

Enabler: Environmental Awareness

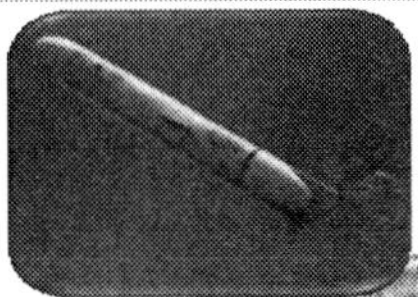

The undersea environment changes significantly after a storm passes requiring constant monitoring.

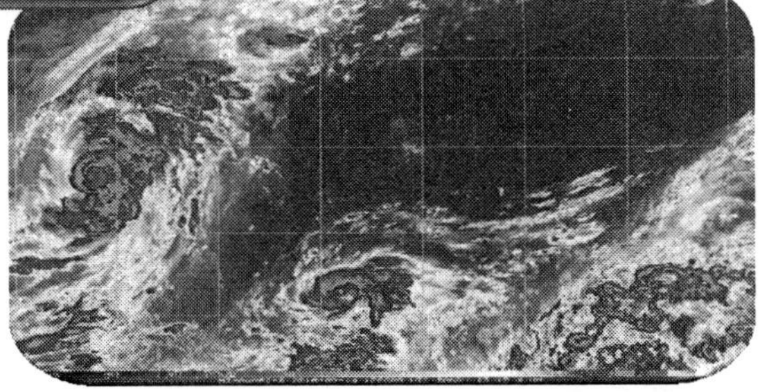

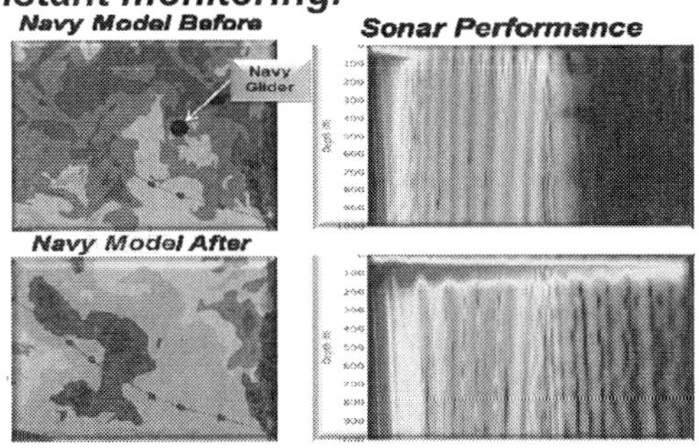

Know the environment or lose the equipment

Naval Oceanography Observe-Predict → Fight-Win

Enabler: Overwatch

Glider Operations Center

Certification Program

CTG 80.7 Command Center

Naval Oceanography has a 24x7x365 Deep Bench

Naval Oceanography *Observe-Predict → Fight-Win*

Enabler: Support Fleet

USNS Maury

T-AGS 60 Class Ships (LARS)

Medium Displacement USV (Seahunter)

Ability to deploy and retrieve unmanned platforms

Naval Oceanography *Observe-Predict → Fight-Win*

Enabler: Maintenance/Warehouse

World class facilities to maintain the Fleet

Naval Oceanography *Observe-Predict → Fight-Win*

Enabler: Certification / Training

Three-Tier Unmanned Maritime Systems Certification Program from the University of Southern Mississippi

- *Tier 1 is a 5-week resident program designed for the UxS "novice"*
    - *Second class is April-May 2018*
- *Tier 2 is a full-semester program for the "journeyman"*
- *Tier 3 provides "expert" level skills in a variety of tailored, specific UxS-related technical and functional areas*

Critical Certification Program for Naval Oceanography

Naval Oceanography *Observe-Predict → Fight-Win*

Enabler: Fleet Transition

Establishing a Department of the Navy and Federal Agency range complex to allow the testing and training of multi-domain, long endurance UxS

Providing authoritative assessment of UxS environmental and hydrographic sensing capabilities under an international "gold standard"

Integrating autonomy, swarming and unmanned system operations with Naval and Marine Corps Command Centers

Bridging the Gap Between Innovation and Operations

Naval Oceanography — *Observe-Predict → Fight-Win*

Enabler: Partnerships

CRADAs
Cooperative Research and Development Agreement

MOA
Memorandum of Agreement

EPAs
Educational Partnership Agreements

MSET

PIAs
Partnership Intermediary Agreement

NOARC

Building our Academic and Industrial Base

Naval Oceanography — *Observe-Predict → Fight-Win*

Future of Navy Oceanography

Distributed data collection, distributed predictive capability, and driving the distributed maritime fight

Naval Oceanography — *Observe-Predict → Fight-Win*

THANK YOU!

Naval Oceanography — *Observe-Predict → Fight-Win*

RED TAIL DOWN: THE SEARCH FOR LAKE HURON'S MISSING TUSKEGEE AIRMEN

The Stockbridge InvenTeam
Feb. 6, 2018

Intro to SHS InvenTeam

How we got involved

Wayne Lusardi, State Maritime Archaeologist visits Stockbridge.

Beginning the research

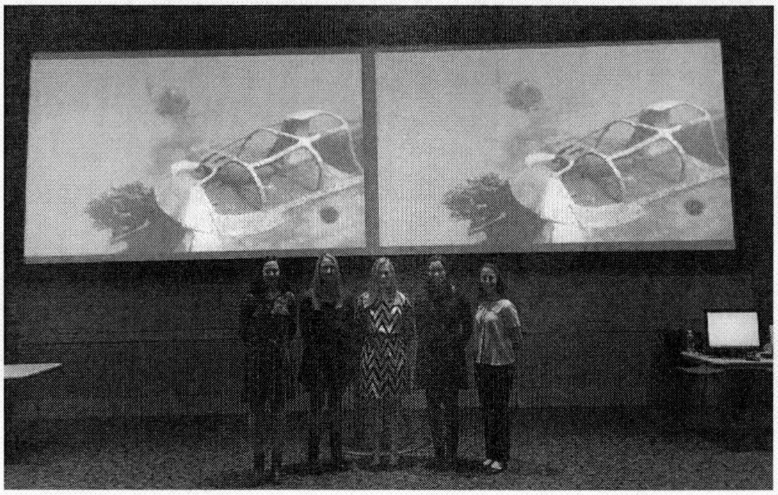

A group of African-American military pilots who fought in World War II

- They were segregated from the white military pilots
- Trained in Tuskegee, Alabama
- Were issued with sometimes faulty P-40 and P-47 planes
- Many pilots crashed during training due to the non-reactive planes

From Alabama
to Michigan

- The Tuskegee Airmen came from Alabama, and moved to Selfridge, Michigan to train for war against the Nazi's for WW2
- Many died during training missions by crashing into the water
- Most who died from training, died because of a plane issue
- Because of segregation still happening, the Tuskegee Airmen were given the "not-so-reliable" planes to train in and use.

Starting from Selfridge

- Selfridge Field was created to train African-American fighter pilots for WW2
- The 1155th, 553rd and 100th fighter squadron have all trained here
- It is located in Selfridge, Michigan
- Thousands of different squadrons have trained here

Heading to Oscoda

- Most Tuskegee Airmen soon after they arrived at Selfridge, moved to Oscoda to practice their gunnery and bombing skills
- Most of the training was conducted at Selfridge Field
- Combat training was all in Oscoda, Michigan

This 1943 official USAAF photo shows the Pilots and Mechanics Of the 100th Fighter Squadron, 332nd Fighter Group, at the Oscoda Army Air Field, Oscoda, Michigan, during combat training,

2nd Lt. Frank H. Moody

- Born in Castle, Oklahoma on December 18, 1921
- Joined the Tuskegee Airmen at age 22 and graduated from the program on February 8, 1944
- He graduated as a 2nd Lt.
- Two months after graduating, he was in a fatal plane crash

Flight Officer Nathaniel P. Rayburg

- Rayburg attended Howard University for 2 years and also the University of Illinois
- During his college years, he was a columnist for the Afro-American Newspaper
- He was apart of the 1155 Fighter Squadron and flew a P-39 Q Airacobra
- On December 12th, 1943, Rayburg crashed in Lake St. Clair during a training mission

2nd Lt. William E. Hill

- Hill flew a P-39 Q Airacobra
- He crashed a few months after he graduated with his class 43-H
- Hill crashed into Lake Huron on November 22, 1943
- His body has still not yet been discovered

2nd Lt. Nathaniel Hill and Lt. Luther Blakeney

- Both were apart of the 100th fighter squadron
- They were piloting a BT-13 Vultee Valiant while observing weather conditions
- On June 16th, 1943, their plane crashed while flying over Oscoda, Michigan into Lake Huron

2Lt. Wilmeth W. Sidat-Singh

- In the late 1930s, Singh played professional sports
- Before joining the Army Air Force, he joined the Metropolitan Police Department of the District of Columbia
- At age 25, he crashed a P-40 Warhawk into Lake Huron near Saginaw Bay on May 9, 1943

Sidat-Singh Athletic Career at Syracuse

Singh on the Syracuse basketball team

Singh on the Syracuse football team

Targets of opportunity

- P-47s
- ***Sgt. Francois Messinger***
- 2LT A. Mikell

Aircraft ID Books

- BentProp Model
- SHS Version

Ship of Opportunity

The Nancy "K"

Coming up with the plan

MATE International ROV Competition

SCUBA and Snorkel

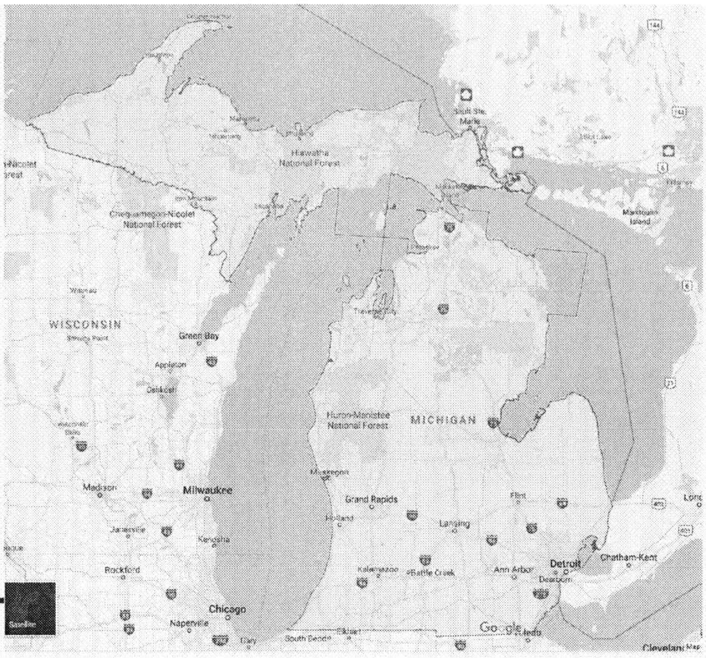

Starting in Traverse Bay

2nd *Leg:* St. Ignace to Presque Isle

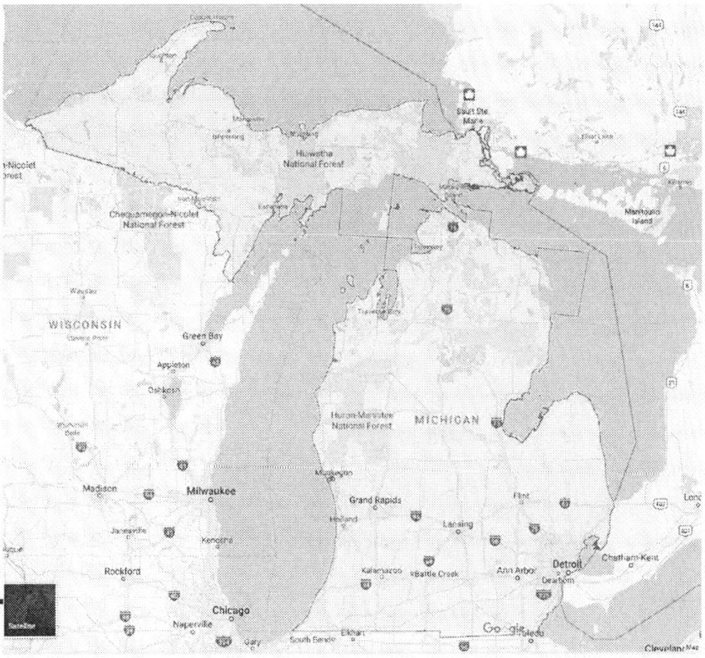
Hiawatha National Forest
Chequamegon-Nicolet National Forest
Manitoulin Island
WISCONSIN
Green Bay
Appleton
Oshkosh
Huron-Manistee National Forest
MICHIGAN
Madison
Milwaukee
Grand Rapids
Flint
Lansing
Janesville
Kenosha
Rockford
Kalamazoo
Battle Creek
Ann Arbor
Detroit
Dearborn
Chatham-Kent
Chicago
Naperville
Gary
South Bend
Elkhart
Cleveland

Riding the Storm Out: Presque Isle to Alpena

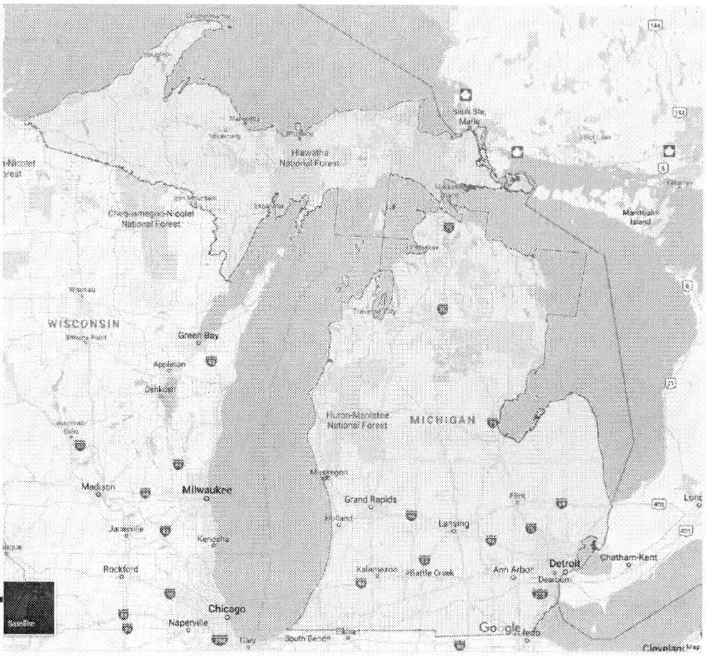
WISCONSIN
Green Bay
Appleton
Oshkosh
Madison
Milwaukee
Janesville
Kenosha
Rockford
Chicago
Naperville
Chequamegon-Nicolet National Forest
Hiawatha National Forest
MICHIGAN
Huron-Manistee National Forest
Muskegon
Grand Rapids
Holland
Lansing
Flint
Kalamazoo
Battle Creek
Ann Arbor
Detroit
Dearborn
Chatham-Kent
South Bend
Google

ROV Practice TBNMS

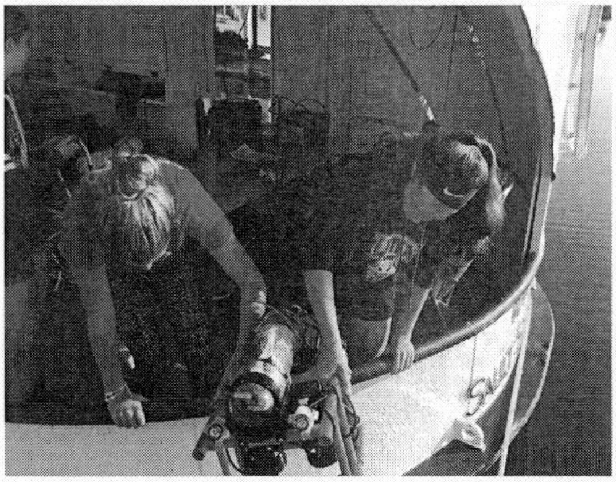

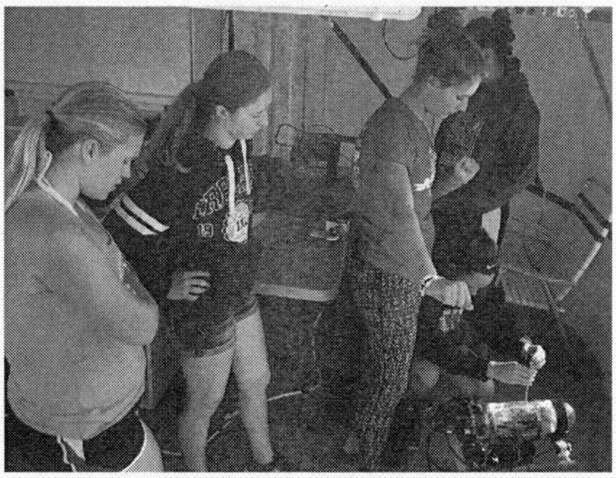

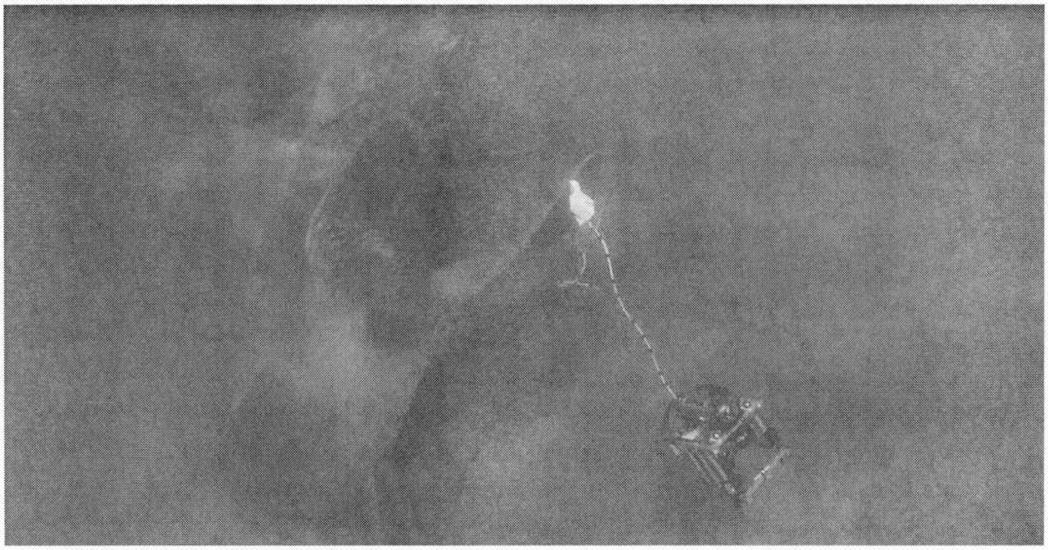

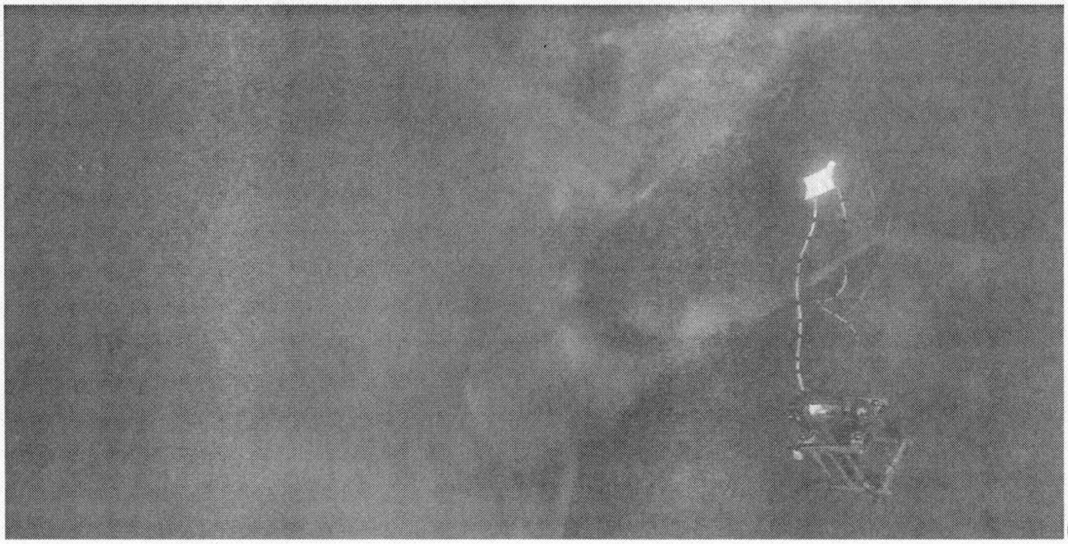

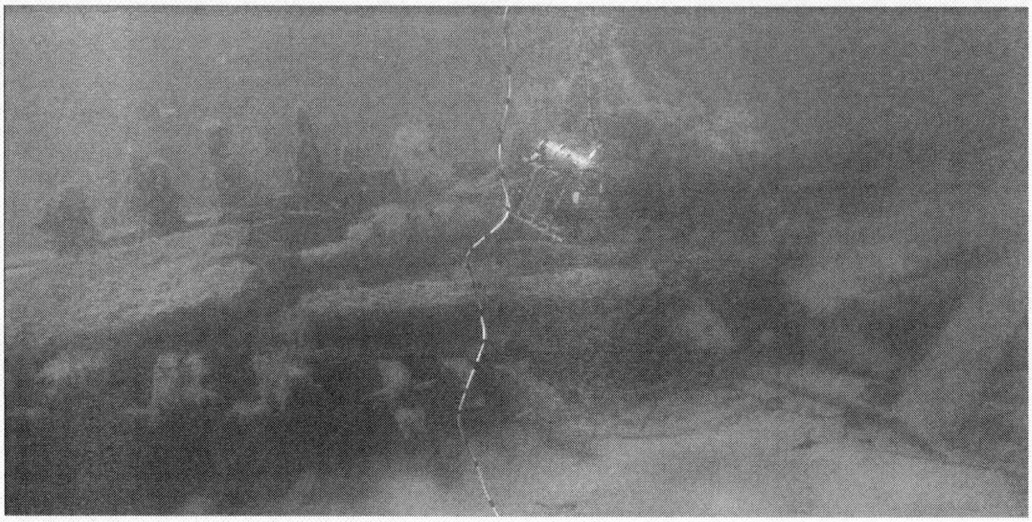

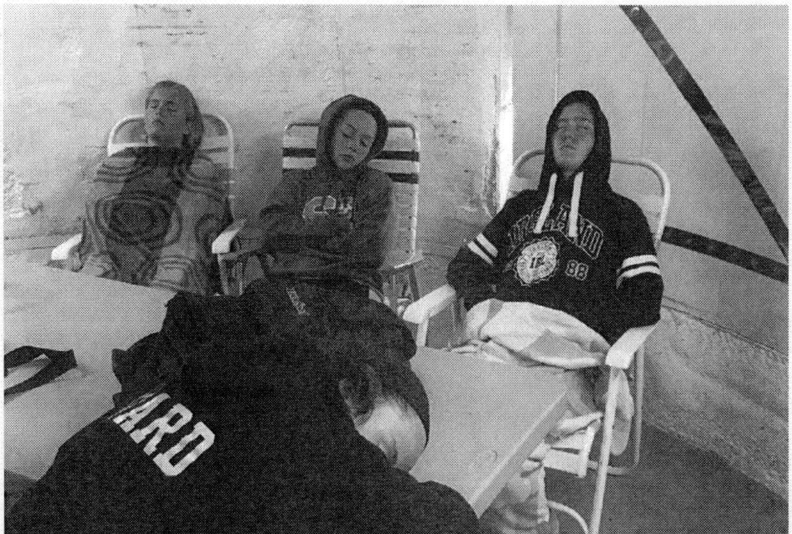

LOWE'S
IGLOO

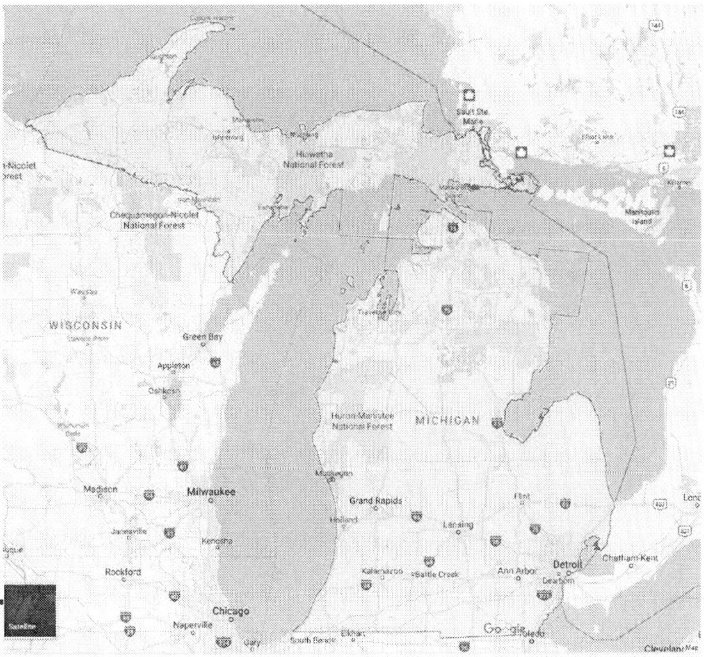
Sault Ste. Marie
Hiawatha National Forest
Chequamegon-Nicolet National Forest
Manitoulin Island
WISCONSIN
Green Bay
Appleton
Oshkosh
MICHIGAN
Huron-Manistee National Forest
Muskegon
Grand Rapids
Holland
Flint
Lansing
Madison
Milwaukee
Janesville
Kenosha
Rockford
Kalamazoo
Battle Creek
Ann Arbor
Detroit
Dearborn
Chatham-Kent
Chicago
Naperville
Gary
South Bend
Elkhart
Google
Toledo

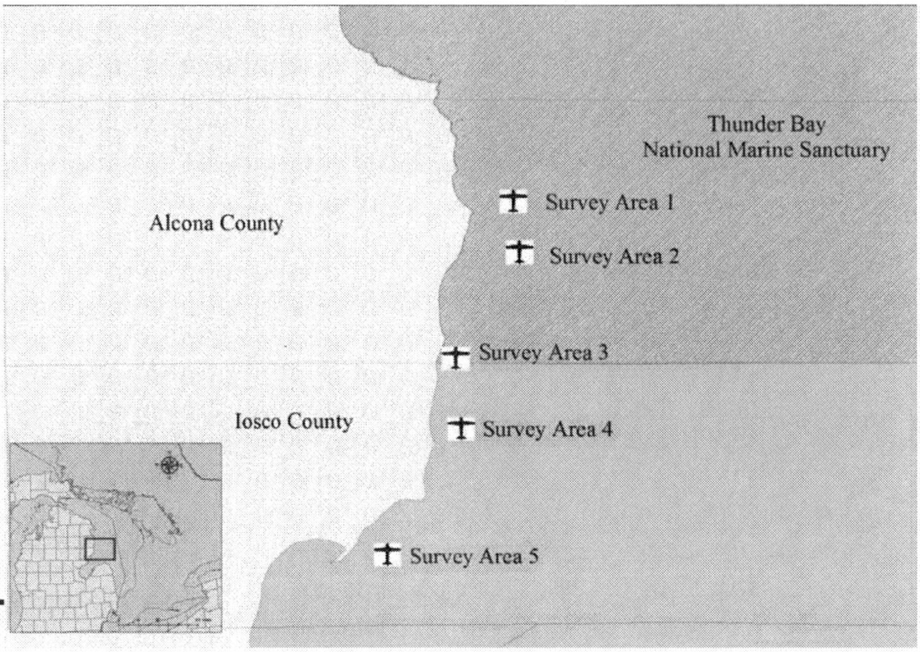

Harrisville to East Tawas

Sonar search

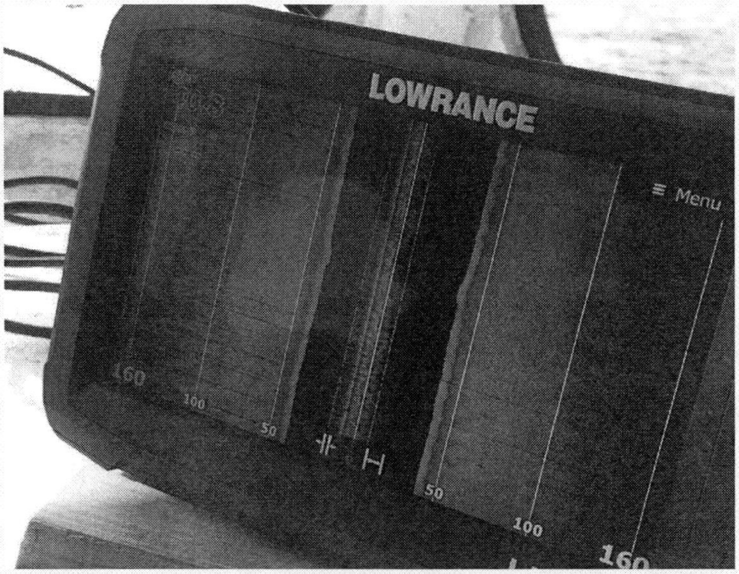

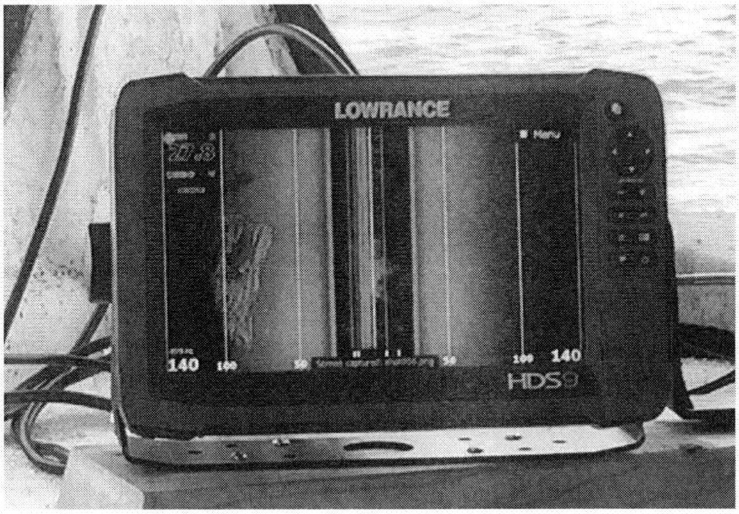

Sadat-Singh:

Where is/was the tower

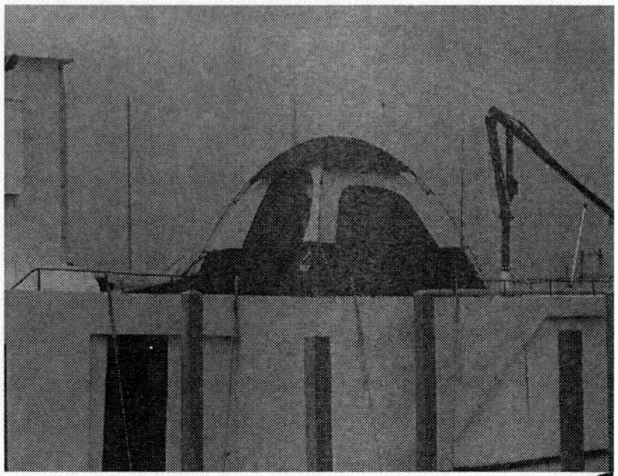

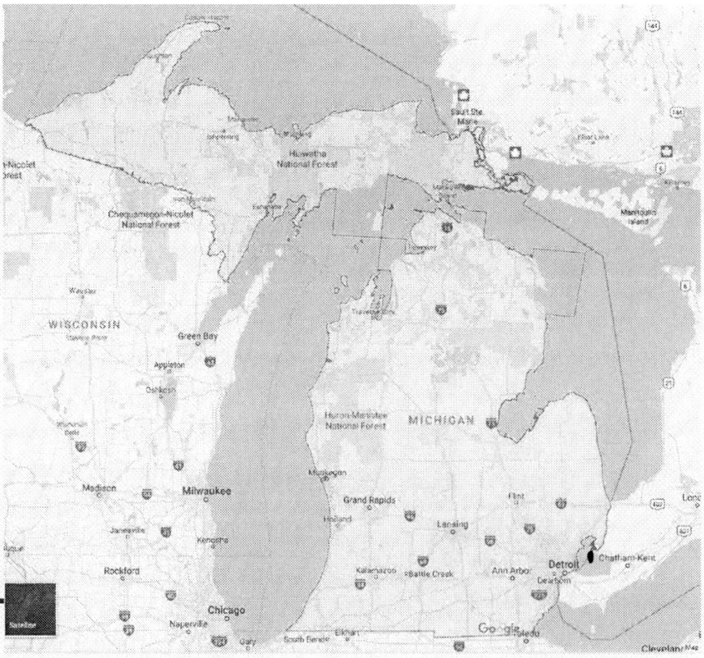
Hiawatha
National Forest
Chequamegon-Nicolet
National Forest
Sault Ste.
Marie
Manitoulin
Island
WISCONSIN
Green Bay
Appleton
Oshkosh
Huron-Manistee
National Forest
MICHIGAN
Madison
Milwaukee
Grand Rapids
Flint
Lansing
Kenosha
Detroit
Chatham-Kent
Rockford
Kalamazoo
Ann Arbor
Dearborn
Chicago
Naperville
Satellite
Google

Blue Water Bridge – across the St. Clair River between Port Huron, Michigan and Sarnia, Ontario

Moody Site

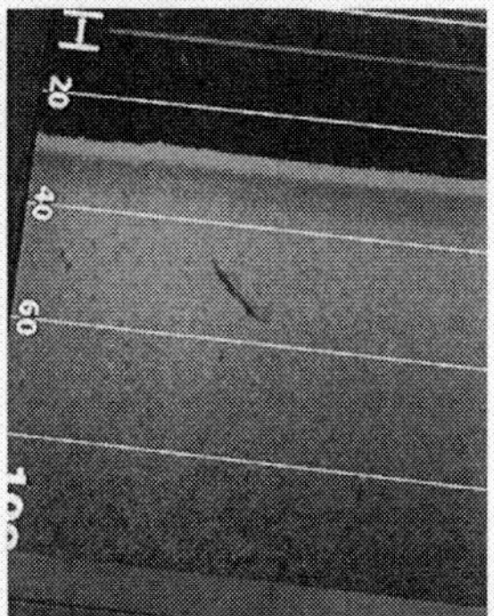

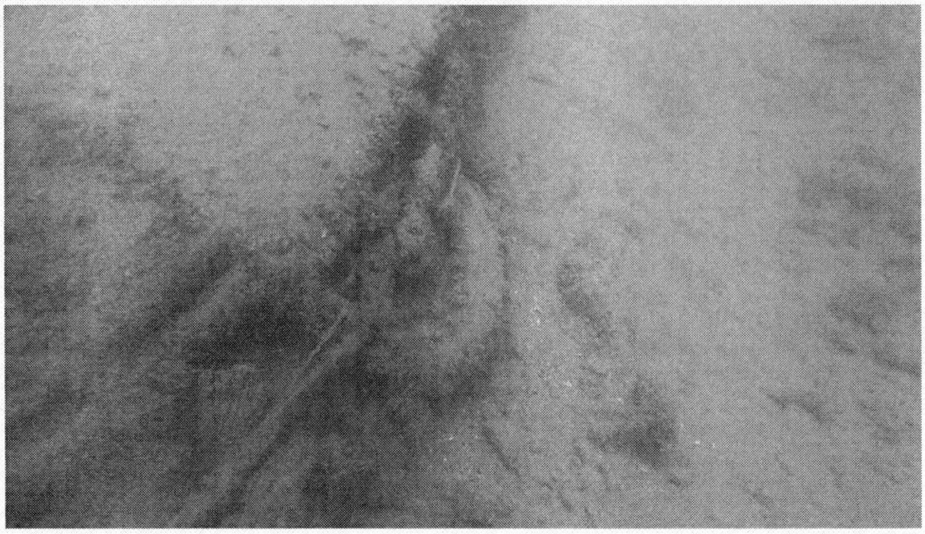

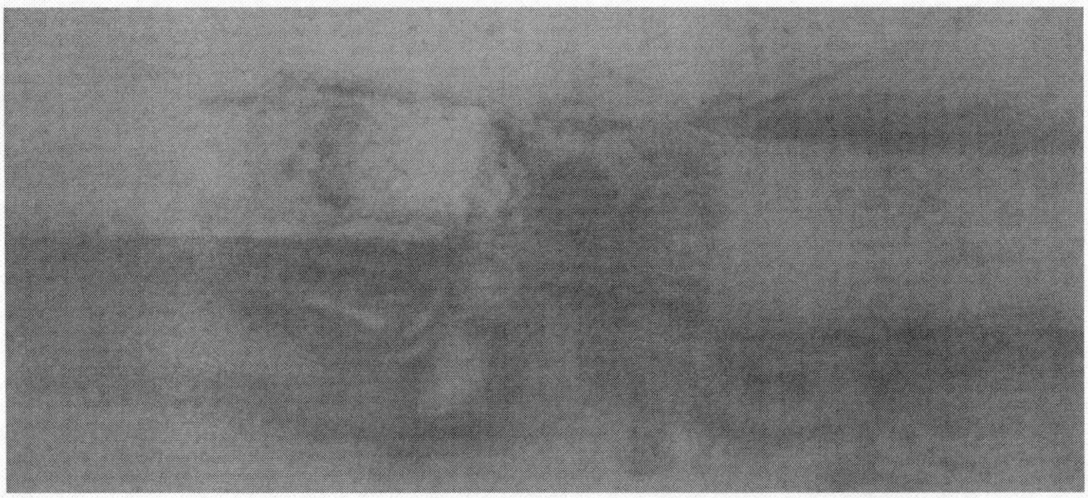

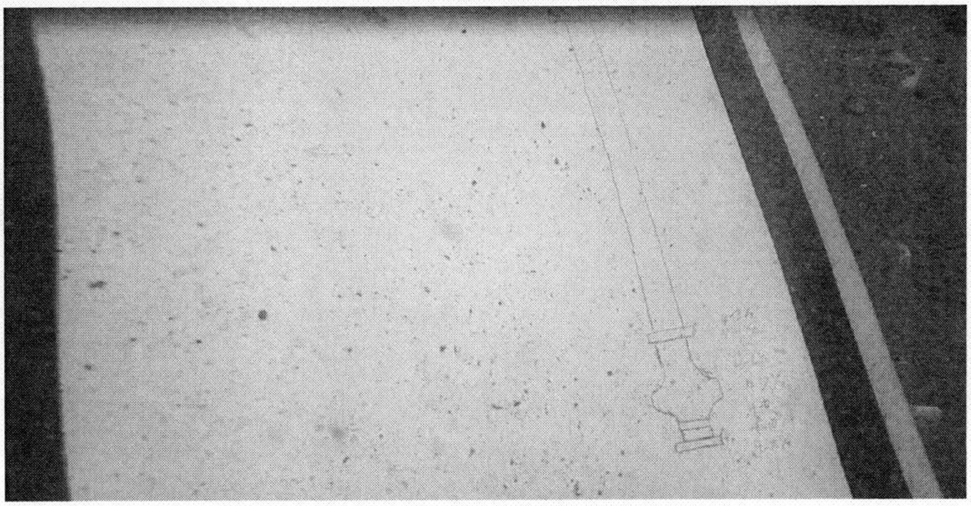

Rayburg Site

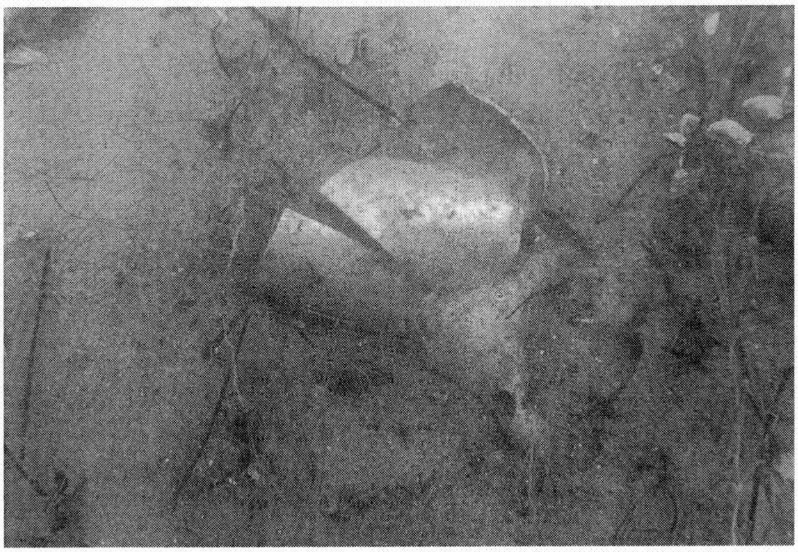

Lessons Learned

- Always use a checklist
- Check everything at least twice
- It's all about teamwork
- Be prepared for anything
- No plan survives first contact
- Hard work pays off

Future Plans

- Summer 2018 Tuskegee II
- American Samoa Fall 2018

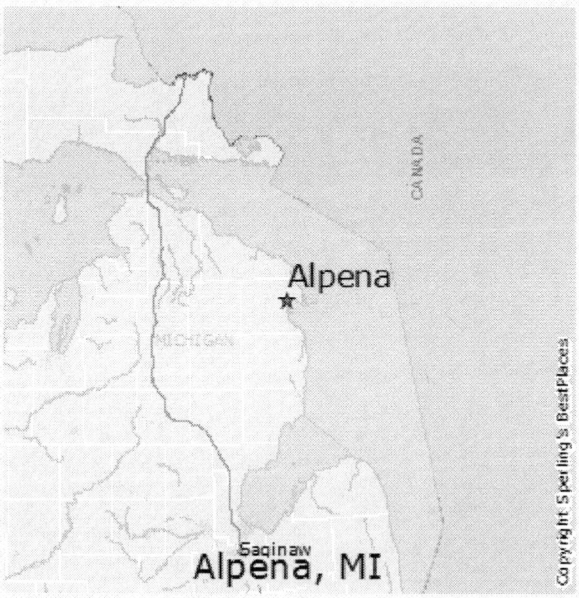

Thank you

- Questions??????????

"Using engineering and 3D printing to Inspire, Engage and Educate, while leaving a positive impact on the world"

Gaining Workplace Experience Through SEAMATE

Dylan Simpson, Maria Camberos Diaz, Nouman Baryal, Savneet Charan
SeaMATE
Monterey Peninsula College

MATE

Background on the Marine Advanced Technology Education (MATE) Center

- The MATE Center was founded in 1997 with funding from the National Science Foundation's (NSF) Advanced Technological Education program. It is currently funded in part by the NSF grants, industry donations, and SeaMATE store sales.
- Headquartered at Monterey Peninsula College, **MATE is an international partnership** of 1,000+ educational institutions, marine industries, and professional societies.
- MATE's mission is to use **marine technology** to inspire and challenge students to learn and **creatively** apply science, technology, engineering, and math (**STEM**) to solving real-world problems in a way that strengthens critical thinking, collaboration, **entrepreneurship**, and **innovation**.

NSF

MATE

SeaMATE is a **social enterprise** that provides community college students with workplace experience while creating products and services that promote engineering and technology education. It is part of the MATE Center and based at Monterey Peninsula College.

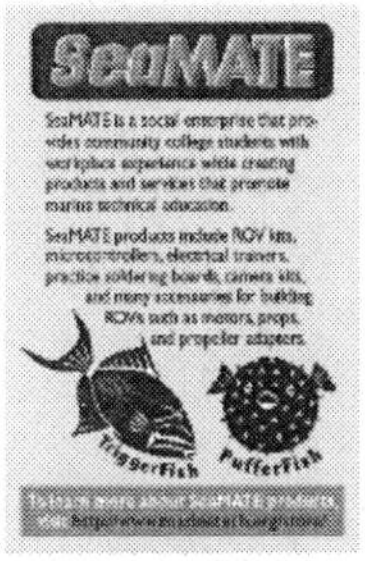

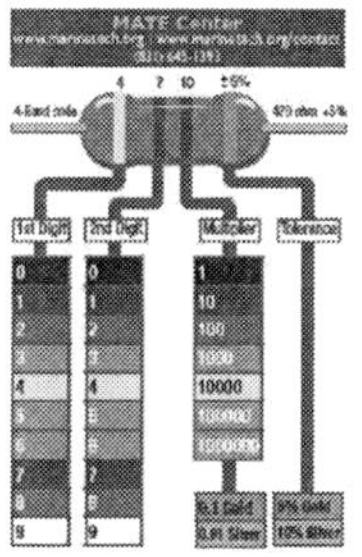

Background on SeaMATE

Created in 2012, to date SeaMATE has…

- Employed 22 Monterey Peninsula College students and provided manufacturing experience to 18 students at Long Beach City College
- Sold 3,000 ROV kits to schools, public aquariums and other educational organizations, families, and "makers"
- Reached $1.5 million in sales
- Seen its students move on to pursue degrees at:
  - Cal Poly San Luis Obispo
  - San Jose State University
  - California State University Monterey Bay
  - University of California Berkeley
  - University of Southern California
- One student employed as an engineer at General Atomics, where she is working on UAVs

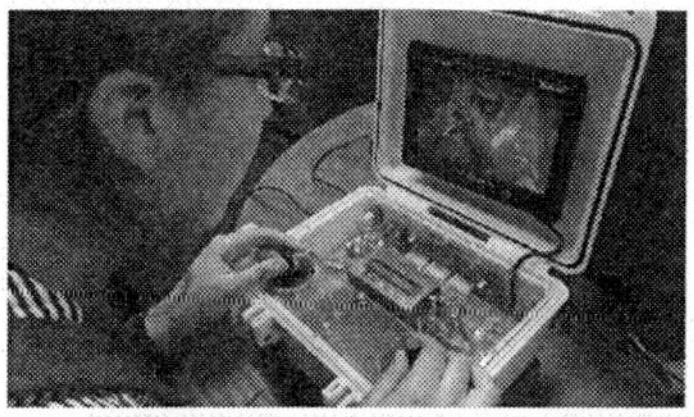

Dylan Simpson

- Studying Computer Science
- I'm interested in environment science, recycling and rock climbing.
- I got involved in SeaMATE in January 2016 when I heard of it in my Marine Biology class.
- My title is SeaMATE Coordinator, I started off in production, moved into inventory and shipping then FQA and now am the overseeing lead of all student work departments.
- I've learned a lot about the many sides of production and running student workshops.
- In the future I plan to go to CSUMB to study environmental engineering.
- My favorite highlight would be workshops up at UCSC.
- Before this job, I didn't consider engineering as a path for me.

Dylan Simpson

MY "AHA" MOMENT

Maria Camberos Diaz

- Studying/interested in:
  - Mechanical Engineering
- Got involved in SeaMATE:
  - Engineering teacher, Tom Rebold about a year ago
- Your progression:
  - Started in production
  - Then assisting in inventory
  - Then taking over shipping
- Job title and description:
  - Lead for Shipping and Inventory
  - Have learned
    - Google Spreadsheets
    - Soldering
    - Using Shopify and shipping procedures
- Plans after SeaMATE:
  - Transferring to a 4 year university
- Highlights of the time so far:
  - Meeting new people
  - Learning new things
  - Participating in the competitions
- SeaMATE's impact on my education and career choices:
  - Has reinforced my love for engineering
  - Broadened my horizon towards more robotics side of engineering

Maria Camberos Diaz

MY "AHA" MOMENT

Nouman Baryal

Studying/interested in:
Studying to Major in Electrical Engineering

Got involved in SeaMATE how/when:

- Fall 2016, and worked as a production assistant
- Workshop -> Inventory

Past Experience:

- Robotics
- S.T.E.M.
- Engineering Clubs and Woodship skills/machinery
- Hobby – soccer
- Year of experience at SeaMATE - in charge of the Production Department

Current Job Title and Learned:

- **Production Lead** - IQC, Assembly, aid production and FQA
- To excel in Google Spreadsheets
- Build, assemble electric wiring & circuitry for ROVs

SeaMATE Highlights:

- Aptos MATE Competition
  - Judging

Plans after SeaMATE and its Impacts:

- Pursue my career at San Jose State University
- Do an internship on one of the vessels and/or related to my field
- SeaMATE impacted me through enforcing of career choices and making me environmentally aware

Nouman Baryal

MY "AHA" MOMENT

Savneet Charan

- Studying/interested in Mechanical Engineering
- I got involved in SeaMATE in the late Spring of 2017. I was working as a peer mentor for my program, and one of my mentees introduced SeaMATE to me.
- I started off as workshop employee; now I am a Technician Specialist.
- I assemble ROV kits and also test their circuitry and operation.
- At SeaMATE I learned how to use Excel spreadsheets, time management, soldering, and different types of circuit components.
- Highlight of my time with SeaMATE - Santa Cruz Outreach.
- SeaMATE has impacted my education and career choices in the following ways:
  - Support
  - Networking
  - Working around my school schedule

Savneet Charan

MY "AHA" MOMENT

WANTED: Community college students looking for technical as well as business experience!

ACKNOWLEDGEMENTS

- The National Science Foundation for funding the MATE Center and helping to start SeaMATE.

NSF

- The Marine Technology Society for funding our travel and experience at Underwater Intervention.

MTS
marine technology society

Questions?

The MATE Center celebrating 20 years!
September 2017

SeaMATE
MATE Center
980 Fremont Street
Monterey, CA 93940
www.marinetech.org/store

Sonardyne
SOUND IN DEPTH

SUBSEA TECHNOLOGY

All-in-one Navigation:

Combining Doppler and Inertial to improve ROV performance

Ioseba Tena
UI 2018

POSITIONING
NAVIGATION
COMMUNICATION
MONITORING
IMAGING

Introduction to Sonardyne

Leading independent provider of underwater acoustic, inertial, optical and sonar technology

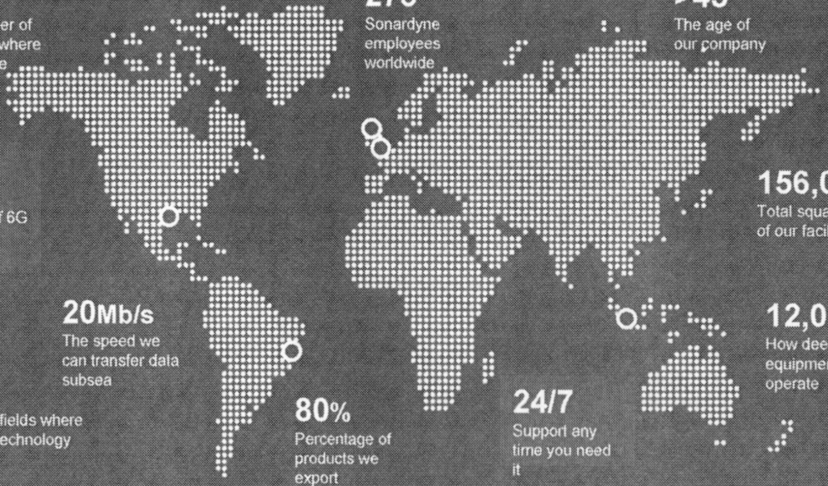

70+ The number of countries where we operate

275 Sonardyne employees worldwide

>45 The age of our company

10,000 Transducer manufactured each year

10mm Positioning accuracy of 6G acoustic technology

156,000 Total square footage of our facilities

20Mb/s The speed we can transfer data subsea

12,000m How deep our equipment can operate

100% Deep water fields where Sonardyne technology is used

80% Percentage of products we export

24/7 Support any time you need it

SPRINT-Nav
All-in-one Subsea Navigation
Sonardyne
SOUND IN DEPTH
ROVs are facing new challenges

SPRINT-Nav
All-in-one Subsea Navigation
Sonardyne
SOUND IN DEPTH
Gravity
Rotation
Image created by Reto Stöckli, Nazmi El Saleous, and Marit Jentoft-Nilsen, NASA GSFC

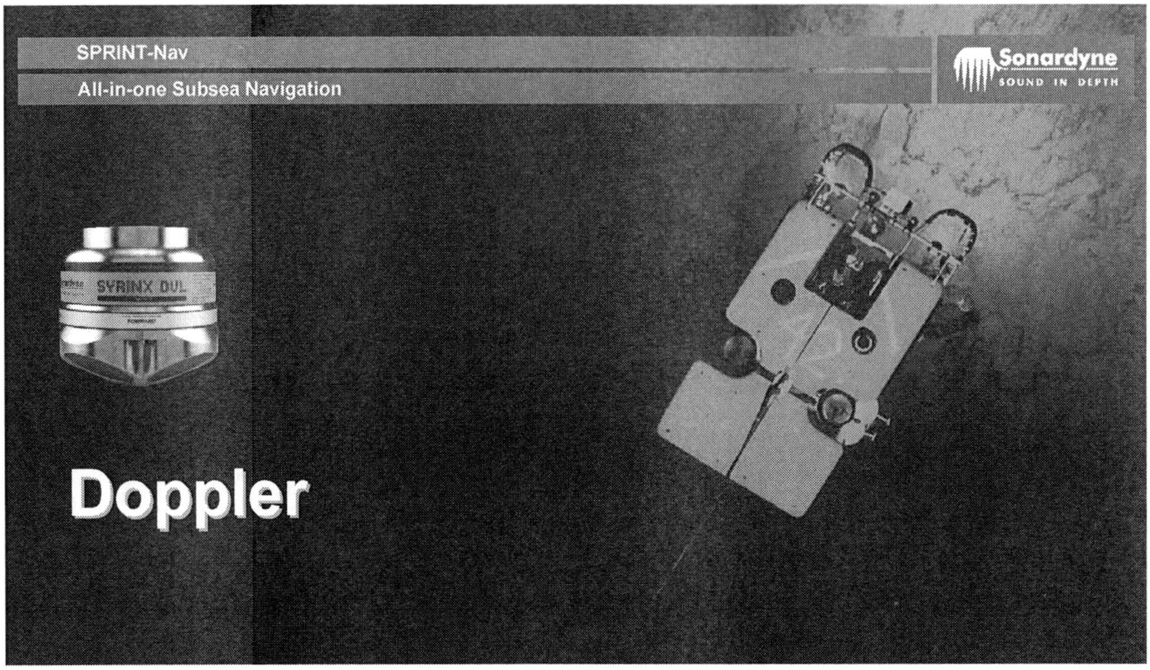
SPRINT-Nav
All-in-one Subsea Navigation
Sonardyne
SOUND IN DEPTH
SYRINX DVL
Doppler

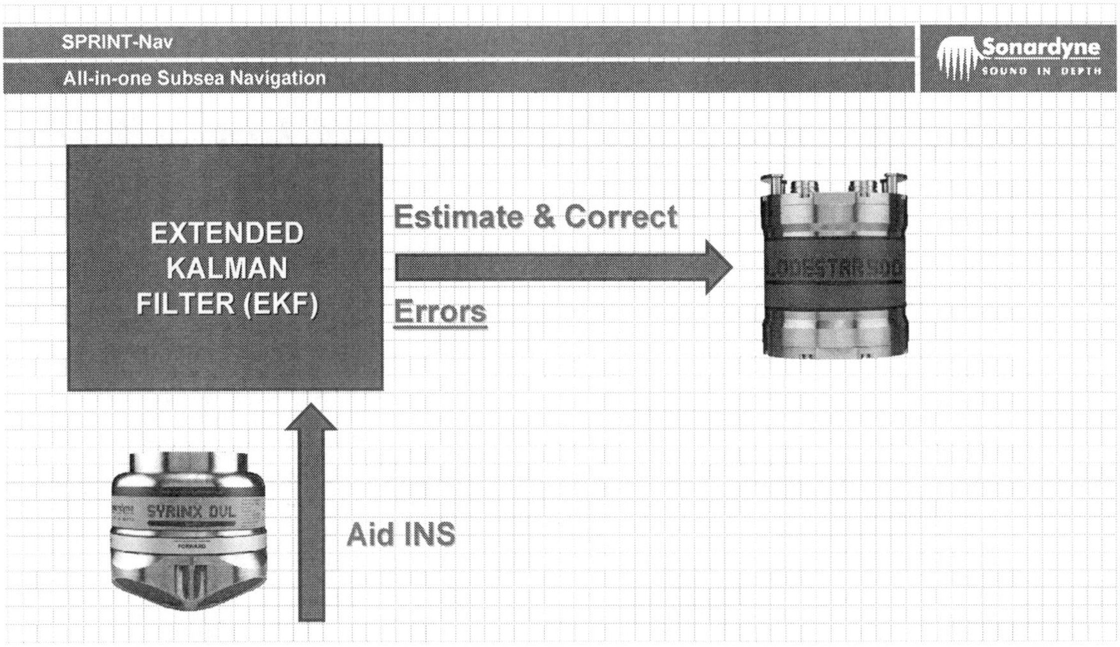
SPRINT-Nav
All-in-one Subsea Navigation
Sonardyne
SOUND IN DEPTH
EXTENDED KALMAN FILTER (EKF)
Estimate & Correct
Errors
SYRINX DVL
Aid INS

SPRINT-Nav
All-in-one Subsea Navigation

Errors in the INS
Errors in the Aiding instruments
Errors calculating offsets

SPRINT-Nav
All-in-one Subsea Navigation

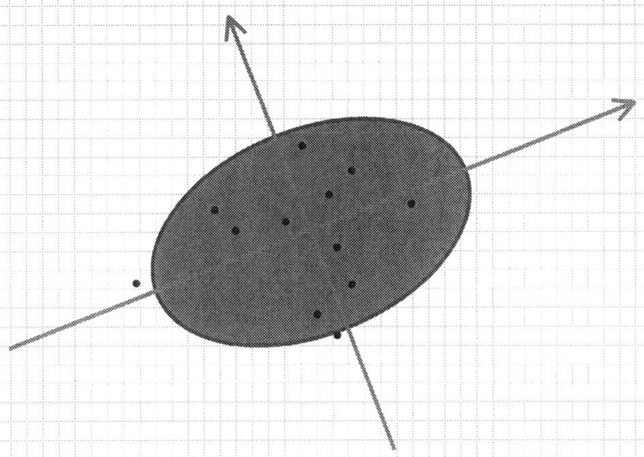

SPRINT-Nav

All-in-one Subsea Navigation

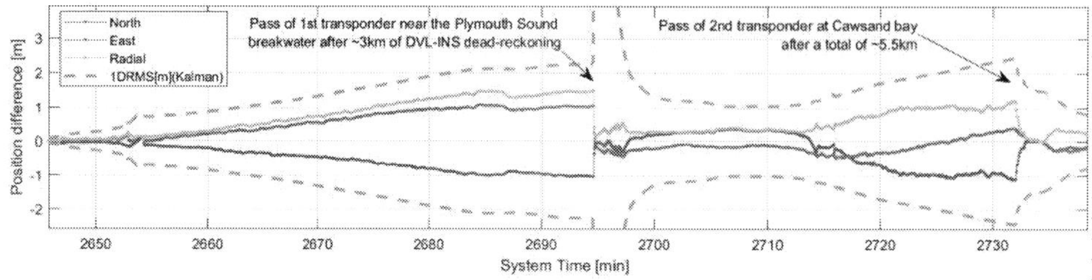

Ellipses vs Time

SPRINT-Nav

All-in-one Subsea Navigation

Tightly Coupled EKF

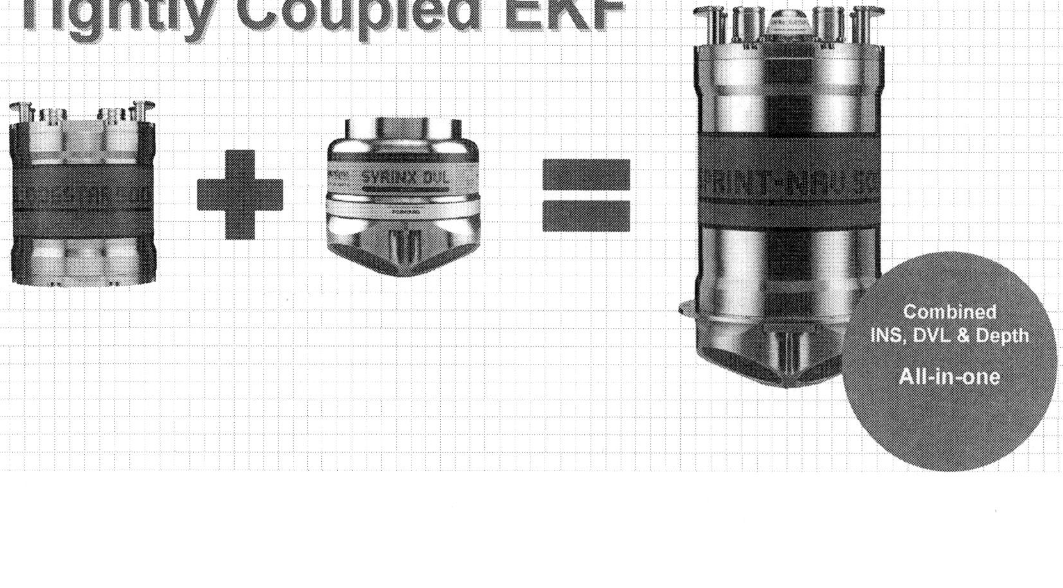

SPRINT-Nav

All-in-one Subsea Navigation

Sonardyne SOUND IN DEPTH

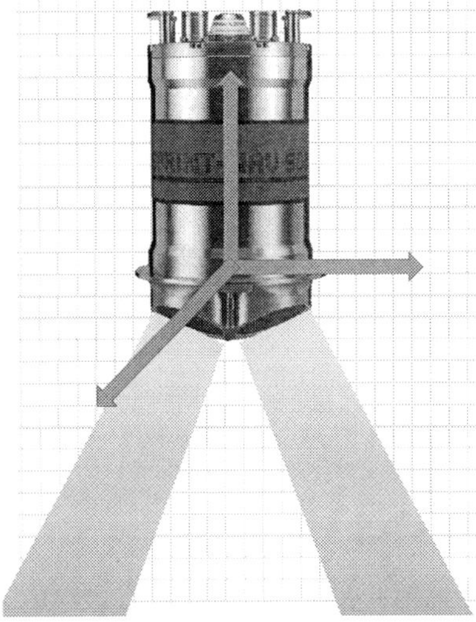

- AHRS alignment for instant INS initialisation
- Mechanically aligned and tightly coupled for improved performance and flexible deployment options
- Access each instrument or simply interface to INS

- Beam level aiding for better performance and robustness even in difficult seabed conditions

SPRINT-Nav

All-in-one Subsea Navigation

Sonardyne SOUND IN DEPTH

- Provides Position, Velocity and Attitude Vectors
- Aided Inertial Navigation runs concurrently with AHRS
- Different performance levels available same size
- 0.01% pressure sensor for complete survey instrument
- INS & DVL continue operating in case of power failure
- No DVL calibration required

- Tightly Coupled Navigation Solution for Best Performance Subsea : we estimate all the errors

SPRINT-Nav
All-in-one Subsea Navigation
Sonardyne
SOUND IN DEPTH
Simple Installation Compact
Shape

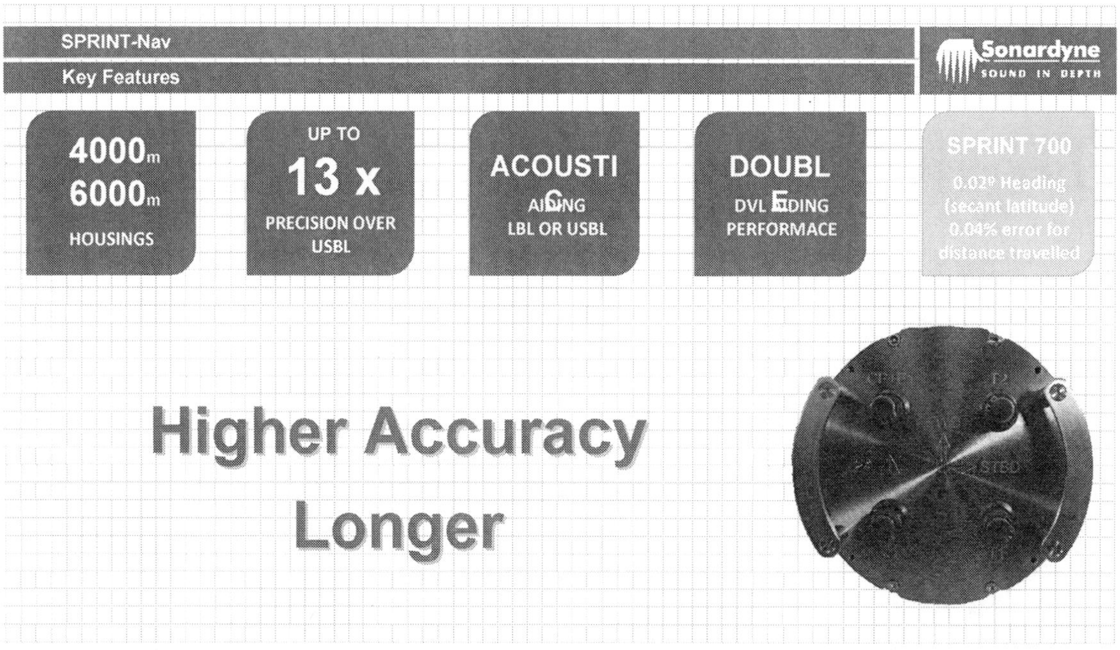
SPRINT-Nav
Key Features
Sonardyne
SOUND IN DEPTH
4000m
6000m
HOUSINGS
UP TO
13 x
PRECISION OVER
USBL
ACOUSTI
C
AIDING
LBL OR USBL
DOUBL
E
DVL AIDING
PERFORMACE
SPRINT 700
0.02º Heading
(secant latitude)
0.04% error for
distance travelled
Higher Accuracy
Longer

SPRINT-Nav
Trials

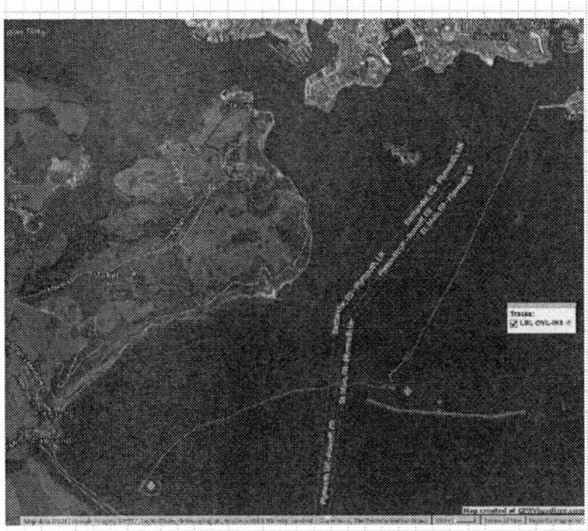

- Long Distance Sparse LBL Trials
- Error of less than 2 m throughout
- Ellipses improve after observing Compatts
- Navigate long distances

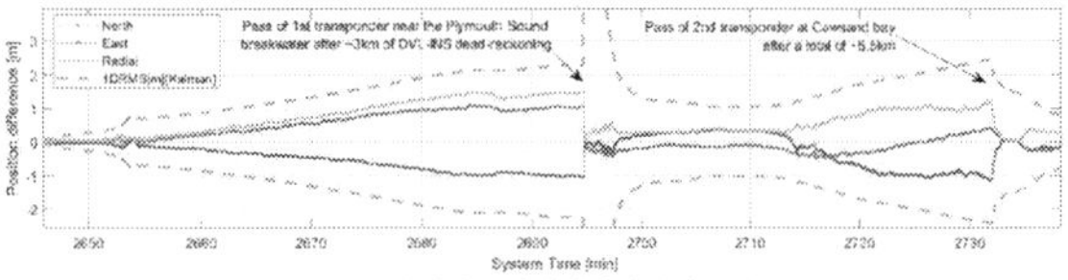

SPRINT-Nav
Navigation for Resident ROV/AUV Concepts

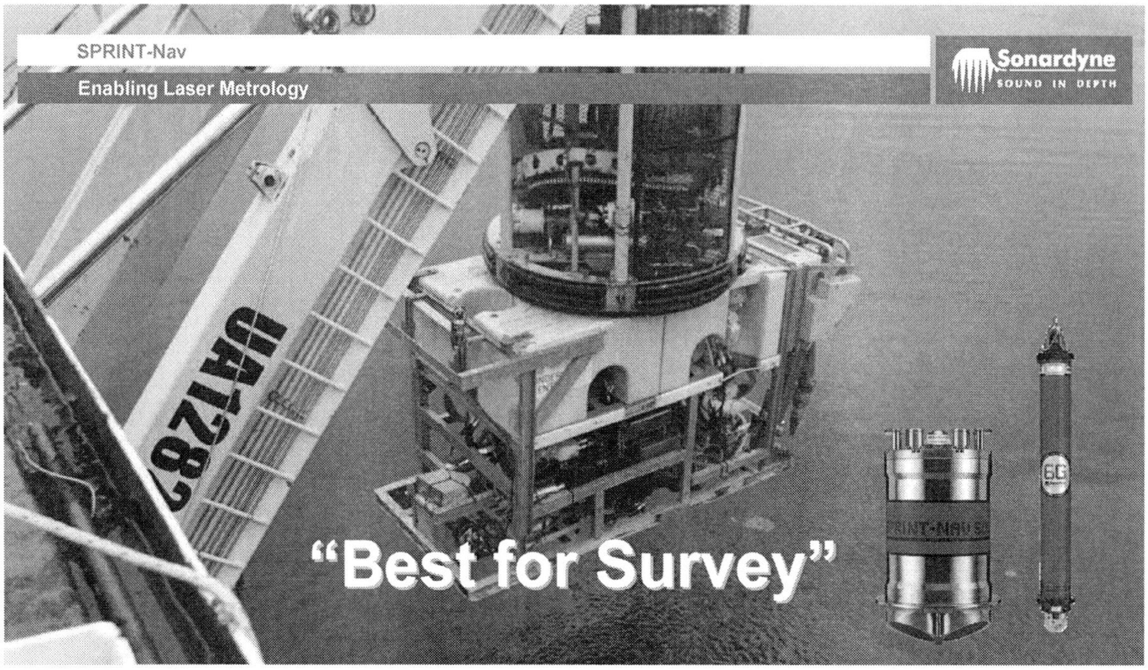

SPRINT-Nav

Enabling Laser Metrology

Sonardyne
SOUND IN DEPTH

Fort William demonstration with DOF Subsea, Seatronics and 2G Robotics

Measurement	Absolute Avg
Horizontal Distance	0.003
Vertical Separation	0.038
Pitch	0.16
Roll	0.39
Heading	0.15

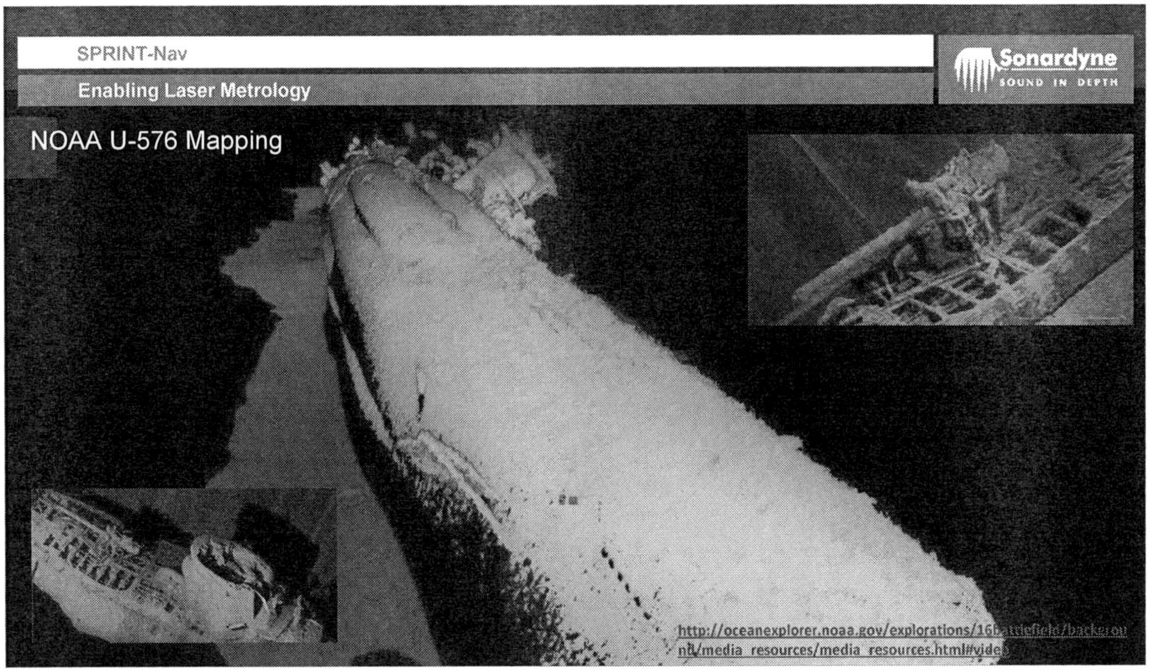
SPRINT-Nav
Enabling Laser Metrology
Sonardyne
SOUND IN DEPTH
NOAA U-576 Mapping
http://oceanexplorer.noaa.gov/explorations/16battlefield/backgrou
nd/media_resources/media_resources.html#video

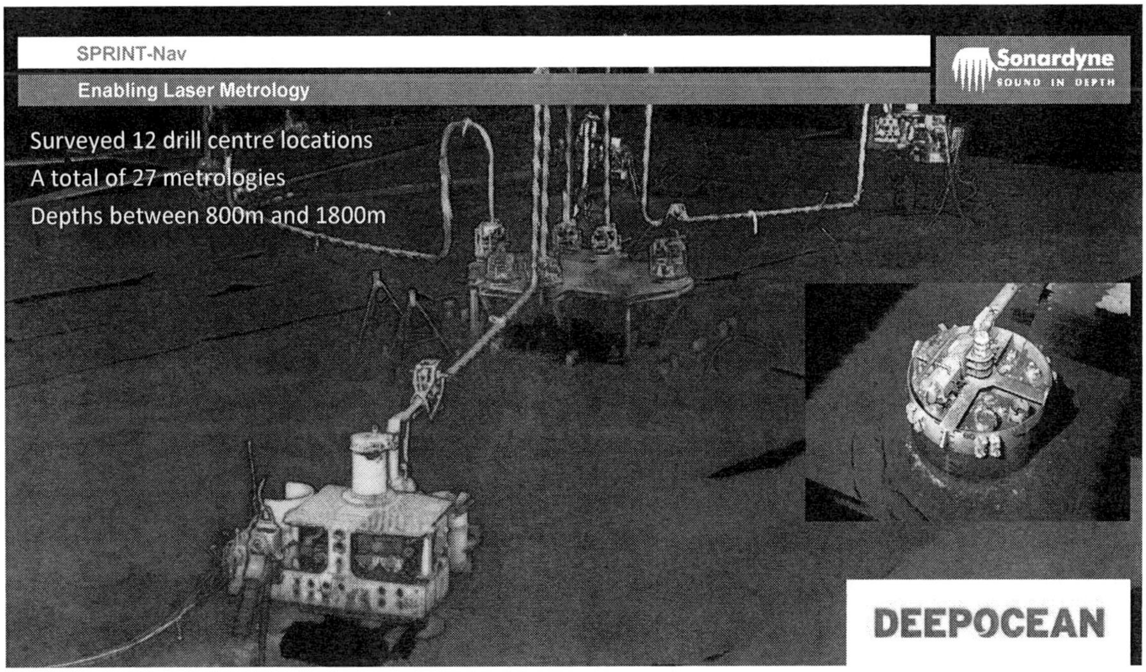
SPRINT-Nav
Enabling Laser Metrology
Sonardyne
SOUND IN DEPTH
Surveyed 12 drill centre locations
A total of 27 metrologies
Depths between 800m and 1800m
DEEPOCEAN

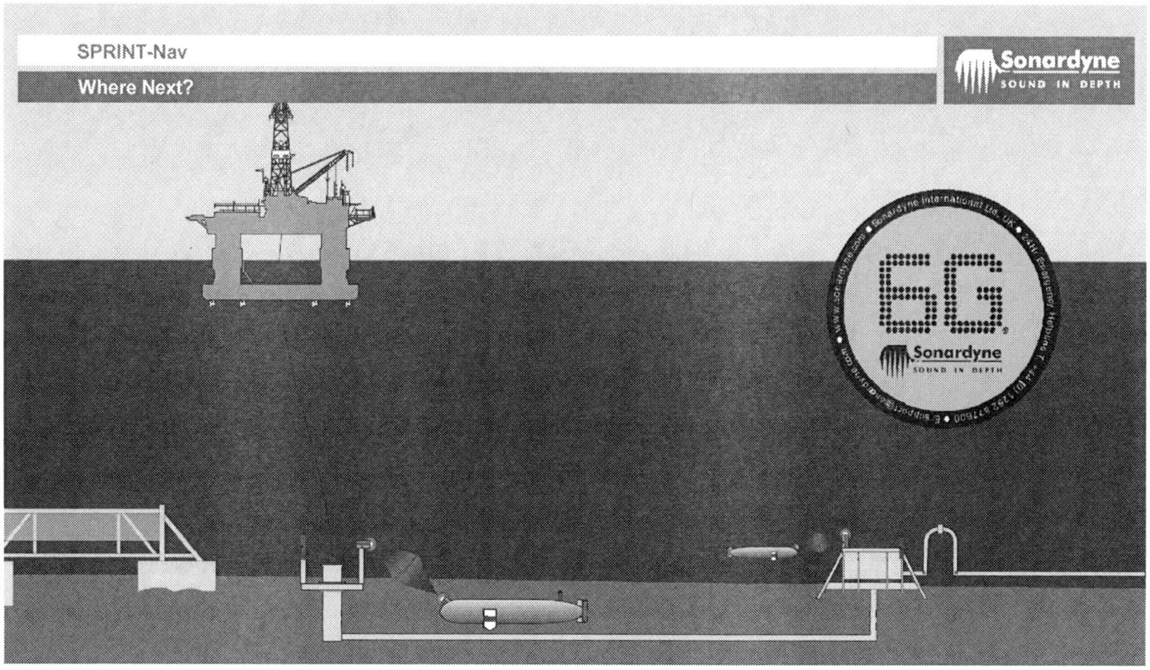
SPRINT-Nav
Where Next?
Sonardyne
SOUND IN DEPTH
6G
Sonardyne
SOUND IN DEPTH

Sonardyne
SOUND IN DEPTH
SUBSEA TECHNOLOGY
Thank you for your time today
Any questions?
SONARDYNE.COM
POSITIONING
NAVIGATION
COMMUNICATION
MONITORING
IMAGING

STEM Outreach Copiah-Lincoln Community College

Kevin McKone, Joseph Crouse, C.T. Smith and Dezirae Katt

π

π

Copiah-Lincoln Community College

- Serves a 7 county region in SW-MS
- 3 Campuses
- 670 Borders
- 3,331Academic and Technical (2016)
- White 53%Black 42%
- 59% Pell Grant
- 68 % receive financial aid

STEM OUTREACH COPIAH-LINCOLN COMMUNITY COLLEGE

Dauphin Island Sealab

Dauphin Island, Al

STEM OUTREACH COPIAH-LINCOLN COMMUNITY COLLEGE

FLL and FTC Kickoff

Wesson, MS

STEM OUTREACH COPIAH-LINCOLN COMMUNITY COLLEGE

First Lego League Championship 2017

Vicksburg, MS

STEM OUTREACH COPIAH-LINCOLN COMMUNITY COLLEGE

Wesson Flea Market

Wesson, MS

STEM OUTREACH COPIAH-LINCOLN COMMUNITY COLLEGE

Super Talk Radio

Brookhaven, MS

STEM OUTREACH COPIAH-LINCOLN COMMUNITY COLLEGE

He-Young Kimm Cup Martial Art Tournament sponsored by the World Han Mu Do Association

Wesson, MS

STEM OUTREACH COPIAH-LINCOLN COMMUNITY COLLEGE

Warren Central High School

Copiah-Lincoln Community College

STEM OUTREACH COPIAH-LINCOLN COMMUNITY COLLEGE

FTC Qualifier

Wesson, MS

STEM OUTREACH COPIAH-LINCOLN COMMUNITY COLLEGE

Nissan Canton

Canton, MS

STEM OUTREACH COPIAH-LINCOLN COMMUNITY COLLEGE

Homecoming

Wesson, MS

STEM OUTREACH COPIAH-LINCOLN COMMUNITY COLLEGE

Georgia Pacific Family Day

Monticello, MS

ELECTRONICS

PRECISION MACHINING

SEAWOLVES ALUMNI

- This Friday and Saturday the Co-Lin Seawolves will be partnering with MATE and Mississippi Robotics
- Our Alumni will be conducting workshops on materials, and CAD
- Alumni at Friday meetings
- Alumni at fund raisers

Thank you the Co-Lin Seawolves!

Follow us on Facebook
https://www.facebook.com/colinseawolves

Providing Insight Onsite

Discovery™ CT Scanning: Mitigating Uncertainty in Pipeline Inspection and Accurately Visualizing and Characterizing Flow Assurance Issues

Jim Bramlett

February 6th, 2018

JM Johnson Matthey
Inspiring science, enhancing life

Objectives

Tracerco

- To demonstrate how the unique advantages of Discovery™ can offer significant value to your subsea inspection projects.
- How?
  - Discuss how operators can quickly identify and accurately determine wall loss integrity flaws.
  - Discuss how operators can pinpoint the exact location, extent and nature of blockages / restrictions.
  - Highlight example applications.
  - Showcase benefits through case studies.

CT - Background

- CT is the acronym for Computed Tomography.
- Penetrating radiation is used to make multiple projections of the sample being inspected from positons all around it.
- From these, a reconstruction algorithm produces a map of attenuation corresponding to a slice through the sample.
  - This is called a tomogram.
- The tomogram is then scaled so that the cross section is displayed in units corresponding to density (g/cc).

Pipeline visualization technology - Discovery™

- Based upon medical Computed Tomography (CT) principles.

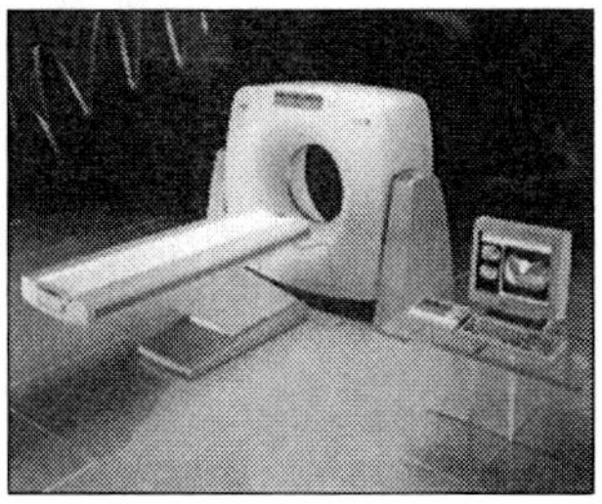

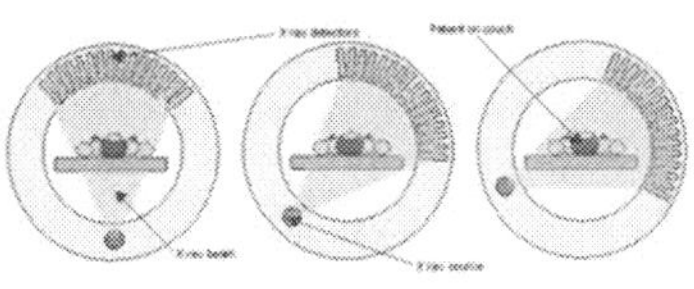

- Due to the significant added value, Tracerco have enabled our customers to use CT technology to save on their subsea pipeline inspection campaigns.

Asset Integrity

Tracerco

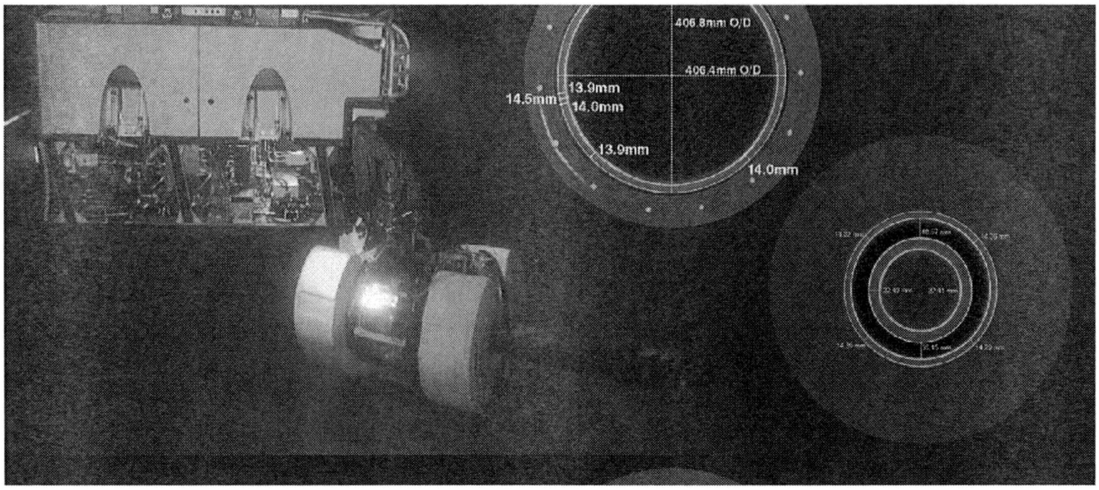

JM Johnson Matthey

The Unique Solution – Discovery™

Tracerco

Why use Discovery™ for Asset Integrity :

- Eliminate the cost and risk of removing coatings subsea.
- Can quickly detect the presence of metal of loss anomalies.
  - Fast screening increases throughput by up to 5 times.
- To assist with ILI inspection operations.
  - Characterize deposits prior to inspection.
  - Validate ILI findings.
- As an alternative to ILI campaigns.
  - Remove the risks of a pipeline cleaning.
- Eliminate interruption to production and normal pipeline operations – therefore no deferment of revenue.
- Reduce operational intervention time so critical decisions can be made immediately by obtaining data online.
- Obtain a complete picture of the integrity of your complex pipeline systems.

➢ Save up to 1/3 on the cost of subsea pipeline inspection.

JM Johnson Matthey

Raw Data (Fast Scanning)

Raw Data after 1 full lap

- Real Time visualization of Raw Projection Data enables the operator to detect presence of metal loss anomalies.
  - Even after just one lap!
- Where an anomaly is detected the scan ...ase the sig... to enable im...
- ...re identified

Raw Data after >20 full laps

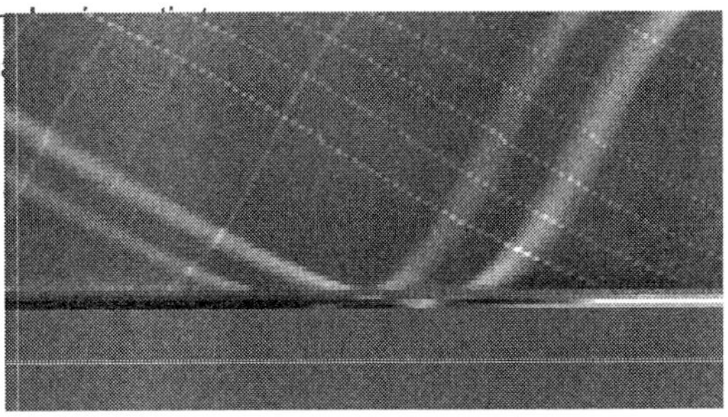

Anomalies detected in raw data → Continue scanning → Sufficient data acquired for image reconstruction

Tomogram

- If an anomaly is observed in the initial projection data then the scanning continues until sufficient data has been acquired to make a well defined tomogram.
- projection data can be easily characterized using the tomogram.
- Tomograms are also available in real time offshore.

Measurement Profiles

- The final inspection report includes 360 degree profiles for the following measurements;
  - Wall thickness
  - Outer radius (on request)
  - Inner radius (on request)
  - Ovality
- Reporting thresholds are also plotted with the profiles.
- Typical tolerance for general wall thickness measurement ±1mm (80% certainty).

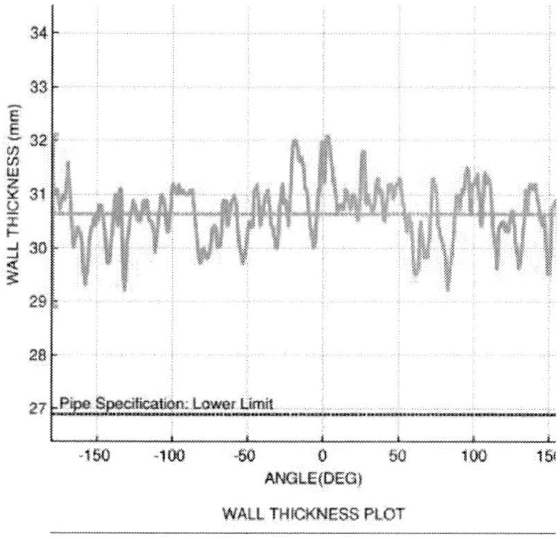

Color Map

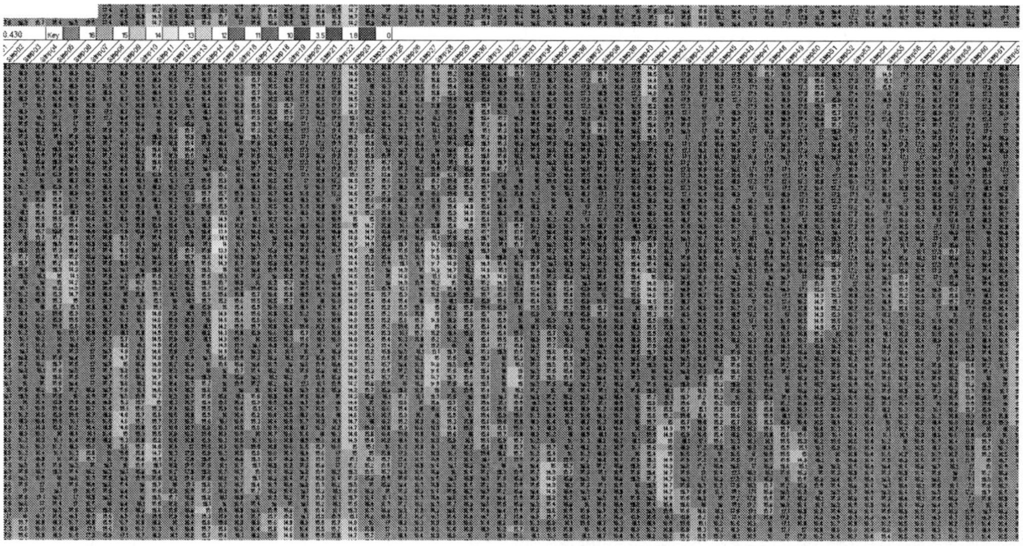

- These profiles are also displayed flat and colored to give an overview of the inspected segment.
- This is an example map of wall thickness measurements – multiple steps over a 1 meter section.

3D Image

Tracerco

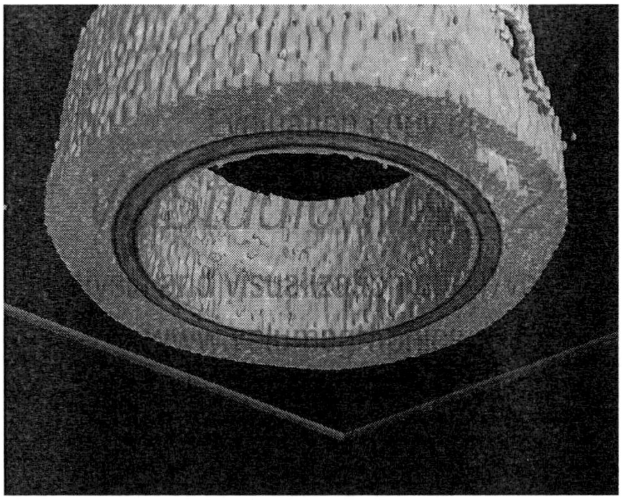

- Upon request multiple images can be stacked to make a 3D part of the inspection location.
- In this example the pipe had a concrete weight coating applied and its roughness is visible to see.

JM Johnson Matth
Inspiring science, enhancing

Characterizing the defect

Tracerco

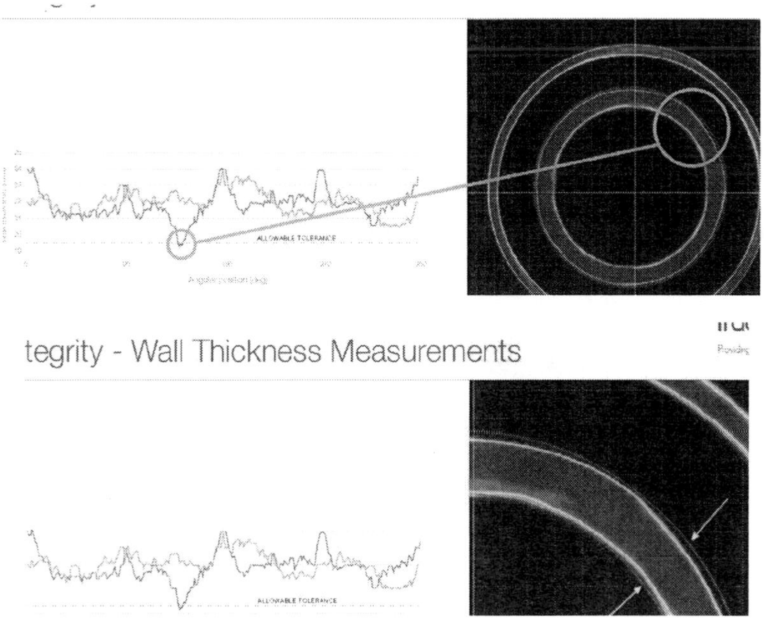

JM Johnson Matth

Flow Assurance & Remediation

Tracerco

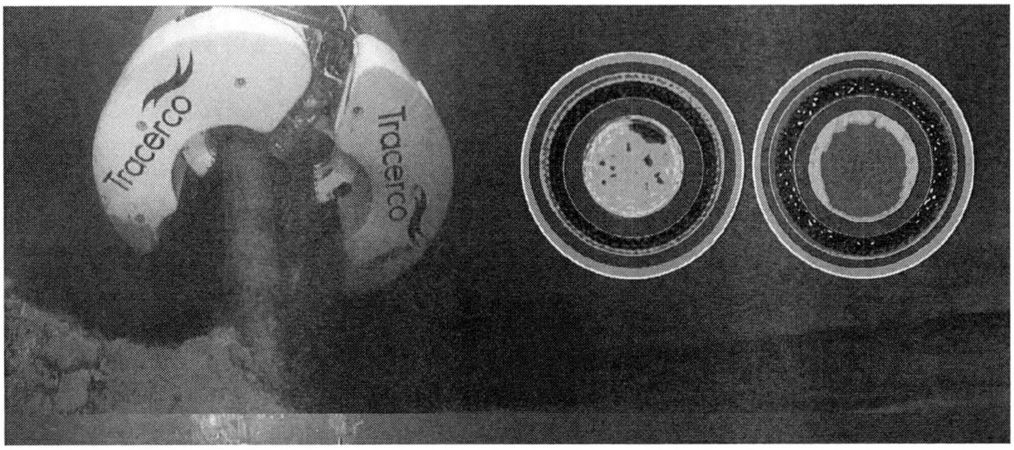

JM Johnson Matthe

Flow Assurance in Pipelines

Tracerco

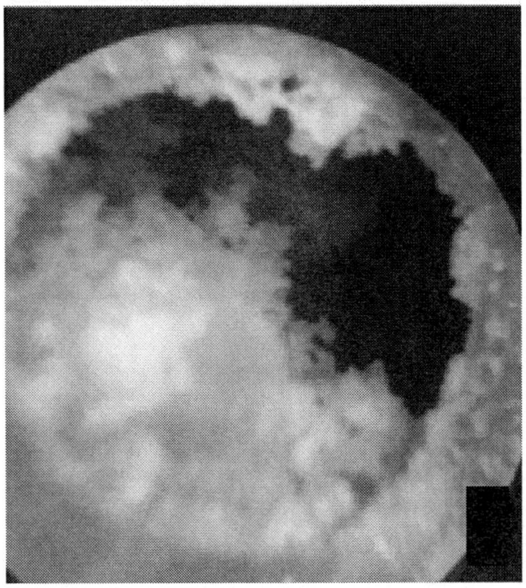

- Operators face the problem of determining the exact location, extent and nature of the build-up and/or blockage in order to take corrective action, without risking damage to the subsea pipeline or further loss of production.

"Research indicates 50% to 80% of remediation attempts fail the first time."

"Incorrect remediation can more than double your overall project costs and damage your asset."

The Unique Solution – Discovery™

Tracerco

Why use Discovery™ for Flow Assurance:

- Can quickly visualize flow assurance issues online and non-intrusively.
  - Enable rapid critical decisions to be made.
- Determine the exact location, extent and nature of blockages in real time.
- Unlock the full production potential of your subsea pipeline and get remediation right first time.
- Take corrective action, without risking damage to the subsea pipeline, interrupting or stopping production.

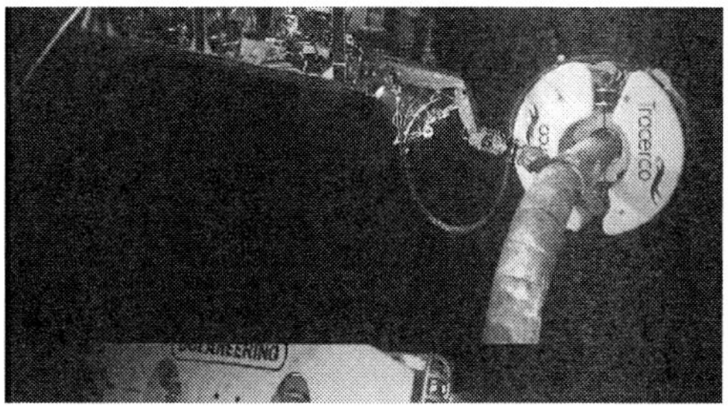

JM Johnson Matthey

Characterize

Tracerco

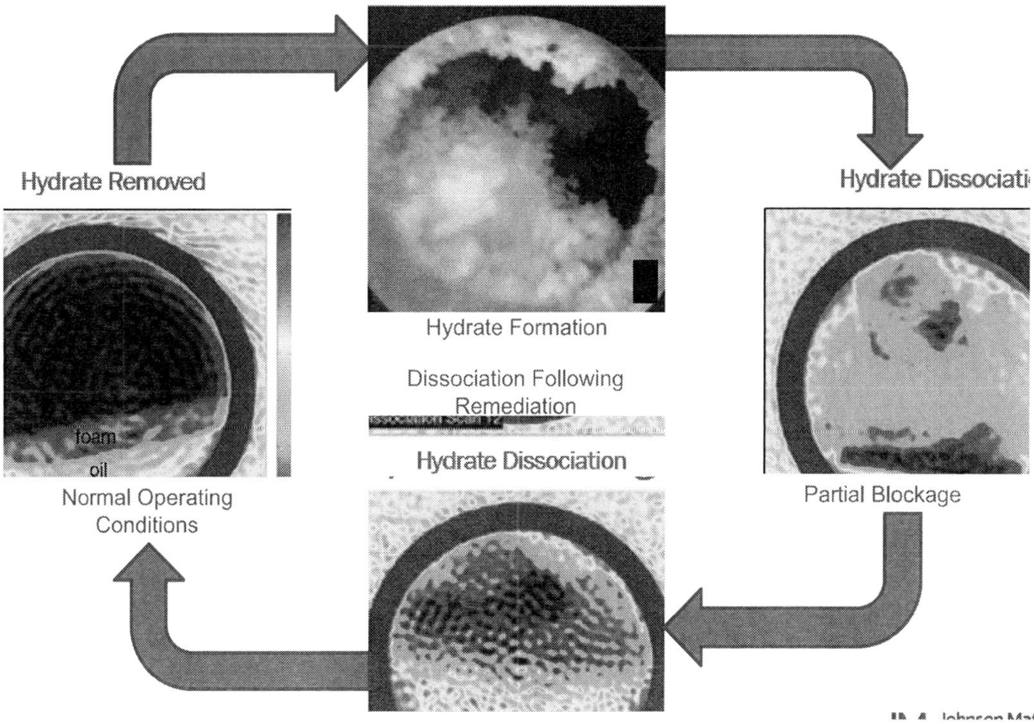

Other Flow Assurance Threats

Tracerco

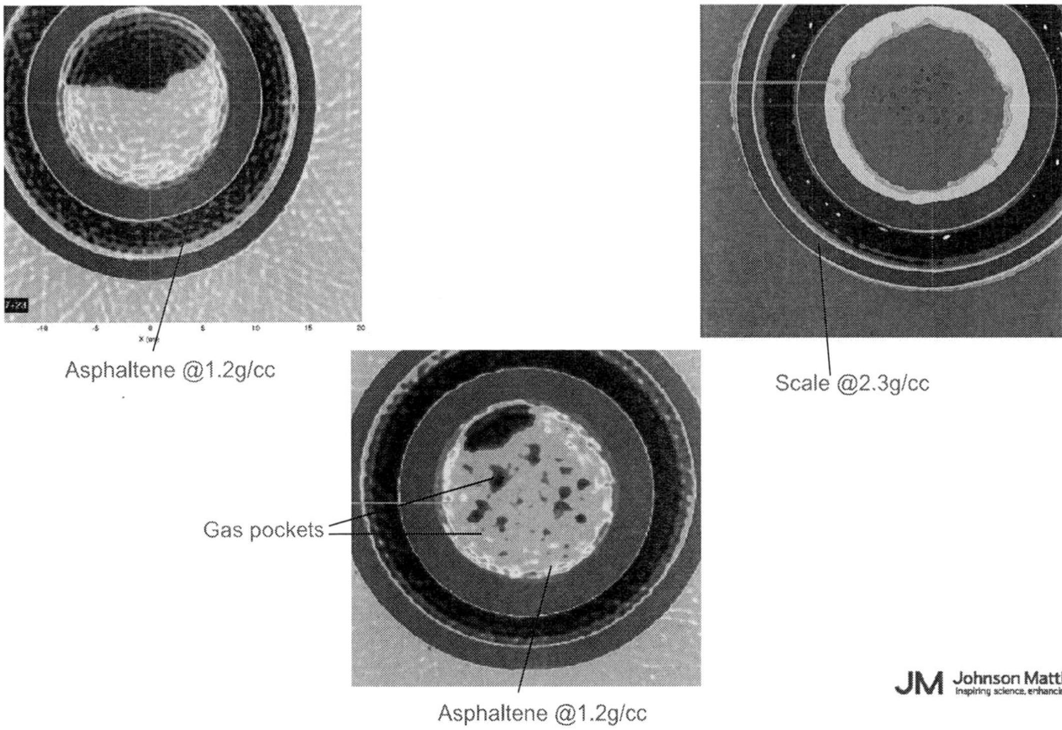

JM Johnson Matthey
Inspiring science, enhancing l

Applications – Coated lines

Tracerco

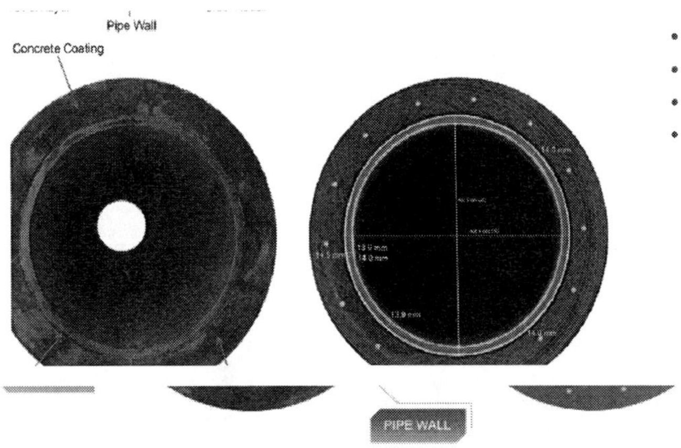

- 360-degree wall thickness map.
- No cost or risk of coating removal.
- Ovality of pipeline under coating.
- Condition of coating and rebars.

Discovery™ scan 360° wall thickness map

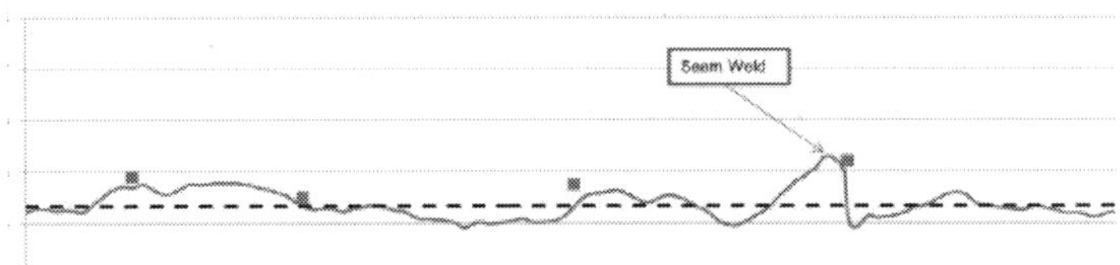

Applications - Flexibles

Tracerco

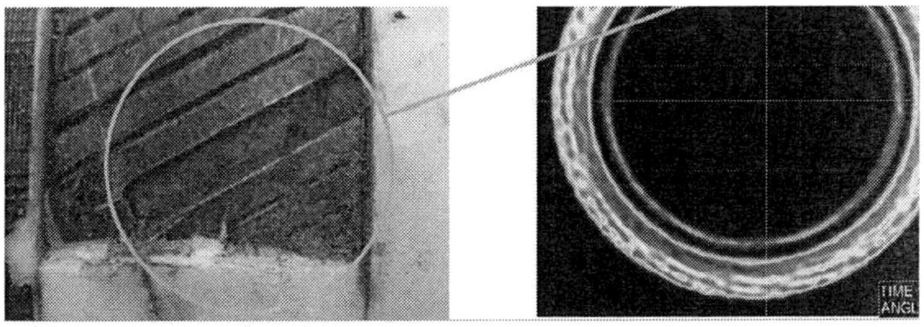

Subsea Technologie

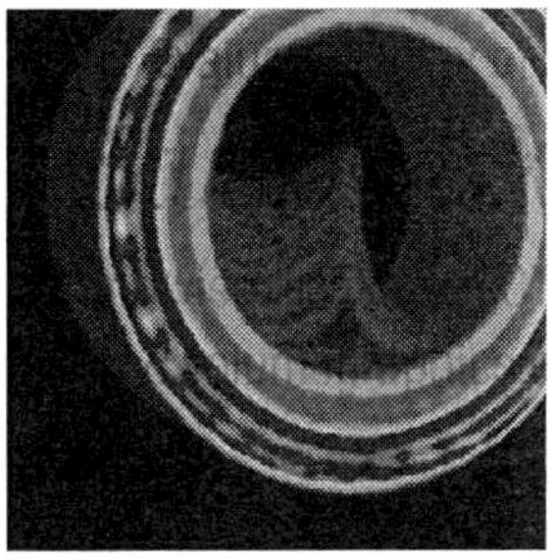

JM Johnson Matth
Inspiring science, enhancing

Applications - Pipe-in-Pipe

Tracerco

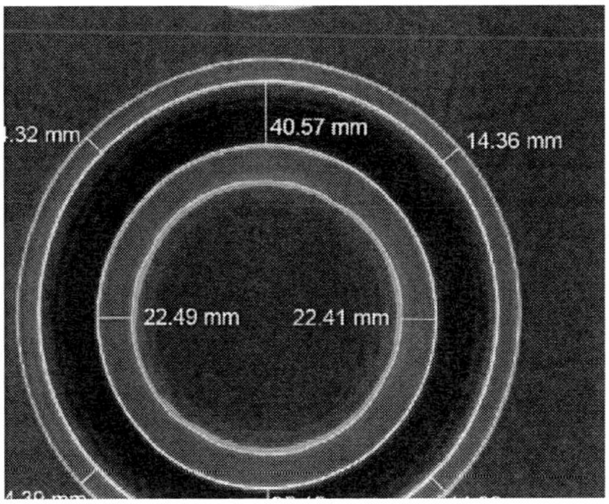

- Scanning through two layers of pipe wall – higher attenuation than a single pipe.
- This is real data taken from an in-service pipe scan.
- Wall thickness accuracy demonstrated to be within Discovery's reported tolerances.

JM Johnson Matth

Applications - Pipe bundles

Tracercc

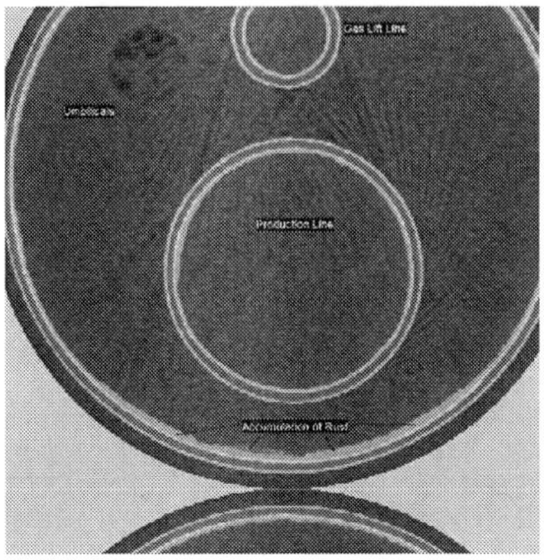

- Outer carrier wall thickness.
- Wall thickness and contents of each inner pipe.
- Able to clearly see the carrier pipe seam weld and an accumulation of rust at the bottom of the pipe.

JM Johnson Matthe

Applications – Piggyback lines

Tracercc

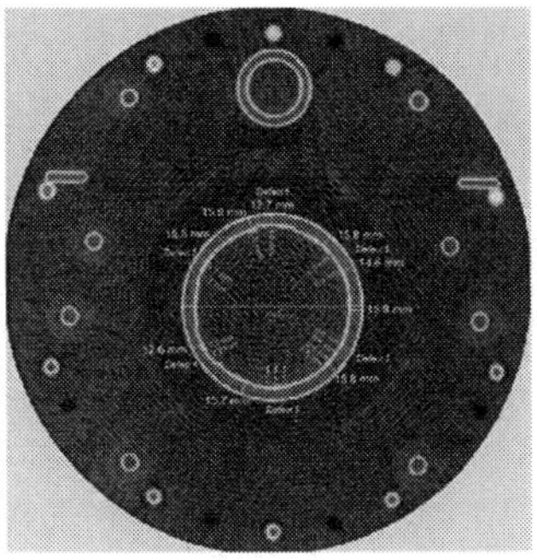

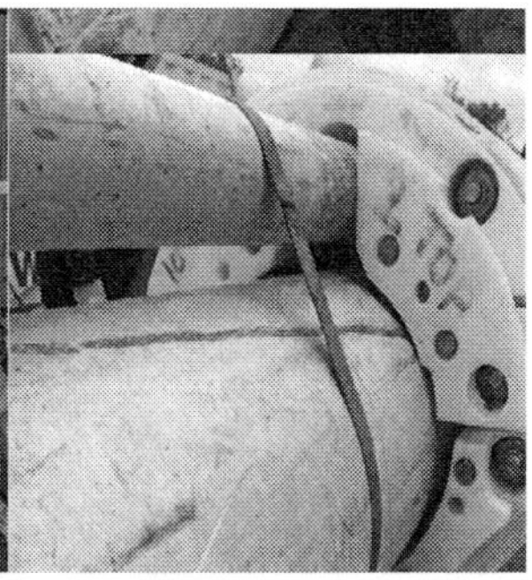

- SoW was inspection of the main export line for lifetime extension purposes.
- Inspection of the piggyback line was not part of the SoW.
- **'You don't know what you don't know'** – water lay-down was detected in sections of the gas lift line.

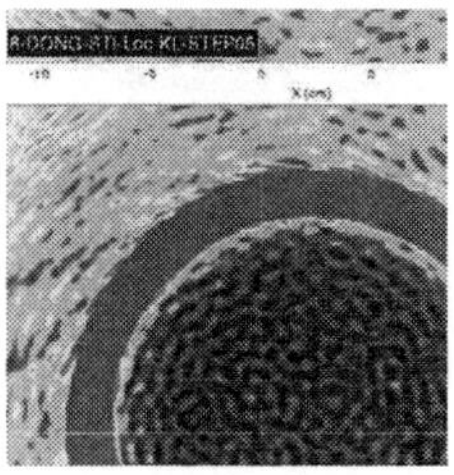

Applications – Risers

Tracercc

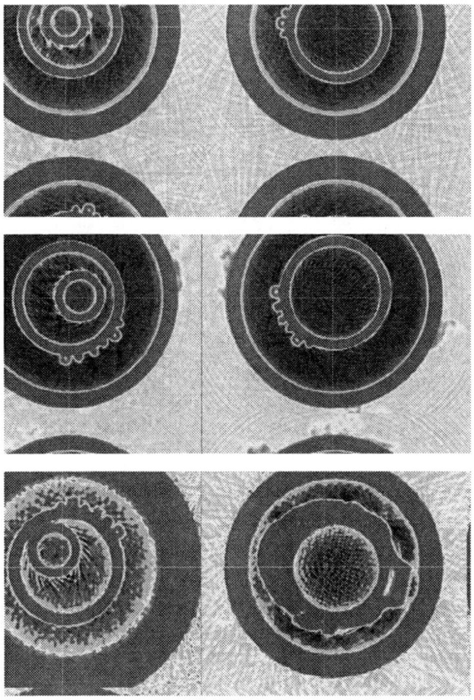

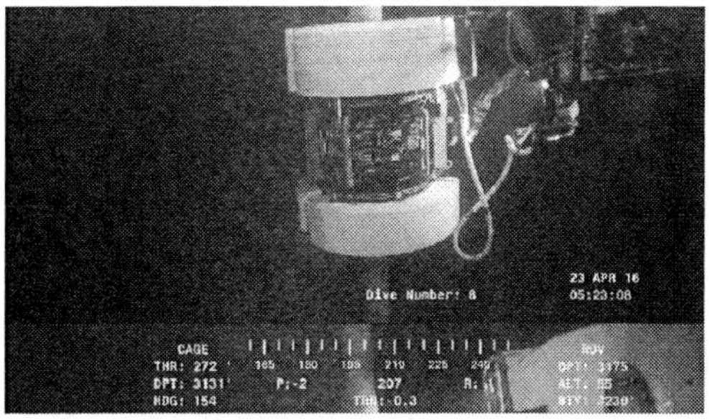

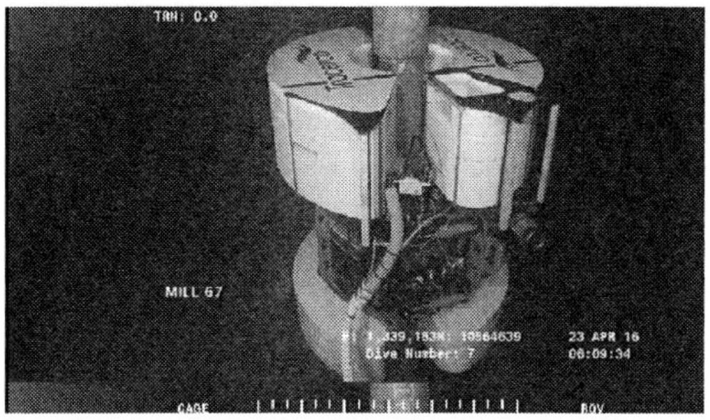

JM Johnson Matthe

Case Study 1

Tracercc

- Subsea CT scanning technology provides vital data in a GoM offshore life extension and flow assurance project.
- Asked to provide a corrosion risk assessment on an existing flowline.
- Subsea CT scanning determined that there was no evidence of accelerated internal corrosion, confirming the integrity of the asset and ensuring it is in a suitable condition for extended life.

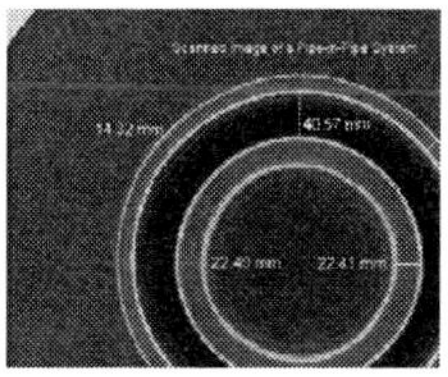

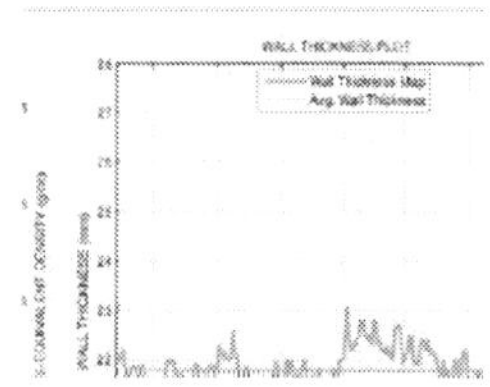

- Outer pipe wall thickness
- Inner pipe wall thickness
- Inner pipe concentricity
- Annulus integrity

Wall Integrity Results

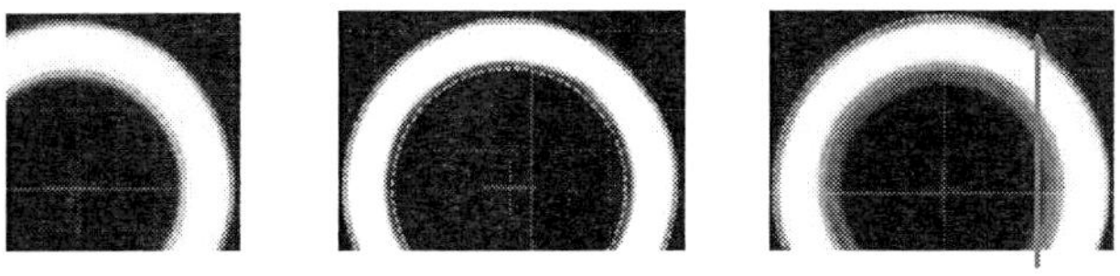

- Build up identified

Case Study 2

Tracerco

- We conducted an extensive inspection campaign in the Gulf of Mexico using Discovery™.
- The campaign delivered:
  - Integrity data that could allow a **lifetime extension of the pipeline.**
  - **$10 million worth of savings** on remediation that would have been ineffectual for the type of build up (restriction identified as scale not wax).
  - **Savings on the inspection campaign.**

"We were already looking at the possibility of using CT scanning technology to inspect pipes for corrosion, then we realized we could use it to see what was inside the pipes too. It was the perfect solution."

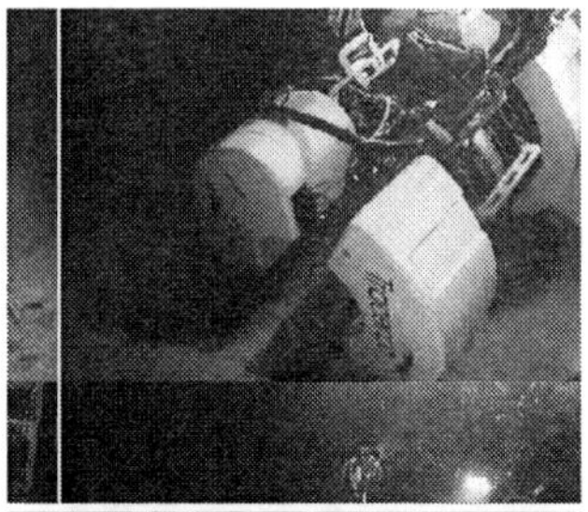

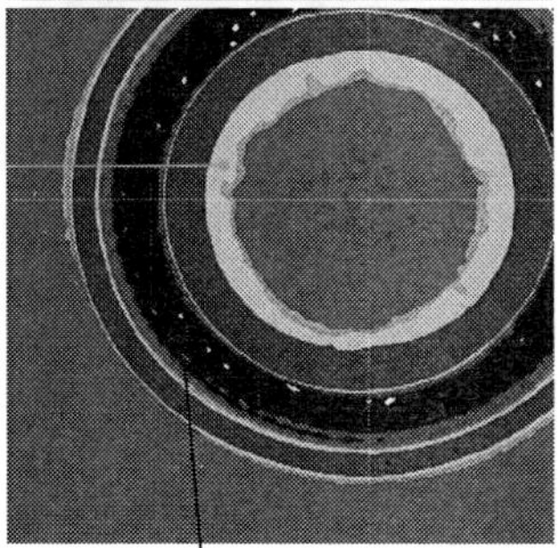

Scale @2.3g/cc

JM Johnson Matthey

Case Study 3

Tracerco

- **Zone 1: 0' -1700'**
  - Liquid full (xylene).
  - Can circulate to manifold.
- **Zone 2: 1701' - 3100'**
  - Liquid to solid (xylene/asphaltene solution).
  - Possible communication to manifold.
- **Zone 3: 3901' - 4560'**
  - Solid asphaltene blockage.
  - No communication.
- **Zone 4: 4561' - 8100'**
  - Asphaltene deposit with trapped gas.
  - Solid and gas mix.
  - No evidence of gas responding to pressure changes at TLP.
- **Zone 5: 8101'**
  - Solid fills approximately 60% of bore.
  - Communication to TLP present.

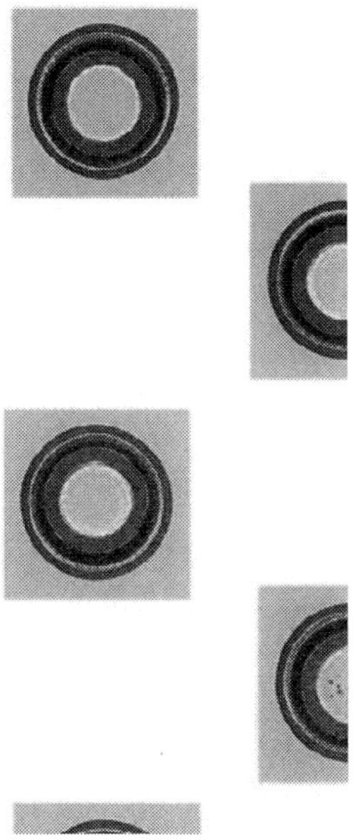

Case Study 4

- Fast Scanning approach utilized to inspect >17m of large diameter gas export line in the Arabian Gulf.
- Very revealing metal loss anomaly identified at the seam weld as shown.

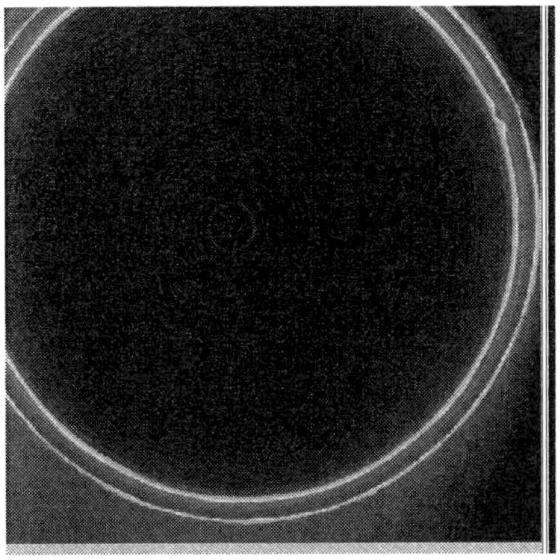

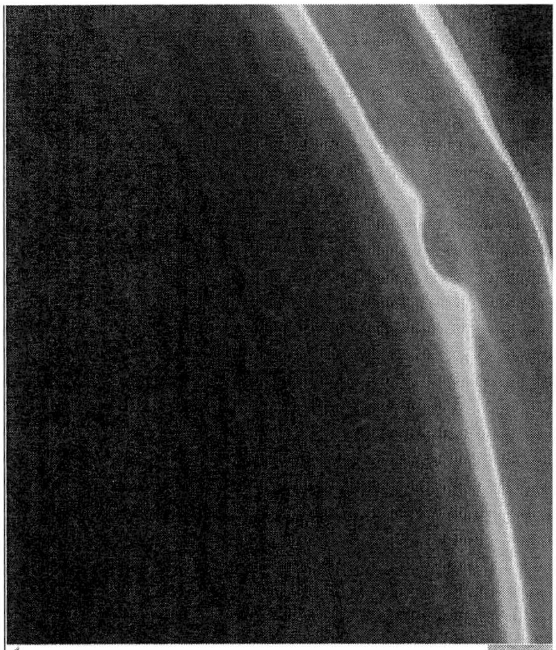

Case Study 4 cont..

- In total, 4 stations were inspected following a typical fast screening approach as shown below.

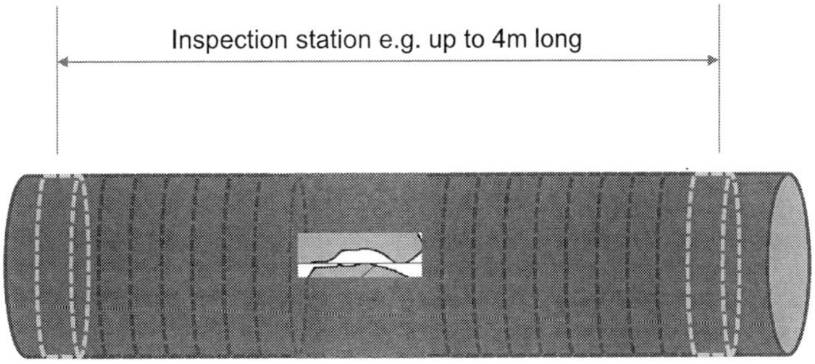

Occasional full duration scans for general wall thickness measurements and deposit assessment

Full duration scans wherever an anomaly is identified

Quick screening scans; instrument quickly steps on if no anomalies are identified

Discovery™ Performance

Tracerco

- Provides an accurate measure of pipe wall thickness through any type of protective coating.
- Can be used on unpiggable and piggable pipelines with any level of complex geometry.
- Fully Operational and Field Proven–TRL 7 (API 17N).
- DNV RP-A203 Certified.
- 3000m Water Depth rated.
- Diameter range up to 26"
  - Larger versions can be manufactured for larger lines.

JM Johnson Matthe
Inspiring science, enhancing l

Summary

Tracerco

- Obtain detailed wall loss integrity data through any coatings and on complex pipelines.
- Eliminate the cost and risk of removing coatings subsea.
- Eliminate interruption to production and normal pipeline operations.
- Visualize flow assurance issues online and non-intrusively, and characterize the type of deposit.
- Reduce costs on remediation campaigns by knowing the exact amount, structure and type of deposits.

Tracerco

Providing Insight Onsite

Questions?

JM Johnson Matthey

Inspiring science, enhancing life

TRELLEBORG

What are we talking about

- What is the current state of the material technology
  - o Issues unique to AUV
  - o Basic definitions
  - o Current State of Technology
    - Material types
    - Manufacturing techniques, cost drivers
  - o Shallow and Deep water buoyancy
    - Syntactic (deep) - General composition and microstructure
      - Microspheres
      - Geriatric aging
    - Unreinforced (shallow) - General composition and microstructure
    - Deep vs Shallow material performance comparison
- Next Generation materials
  - o Integral buoyancy – guards; connectors; fairings

2

TRELLEBORG

Challenges in AUV Buoyancy

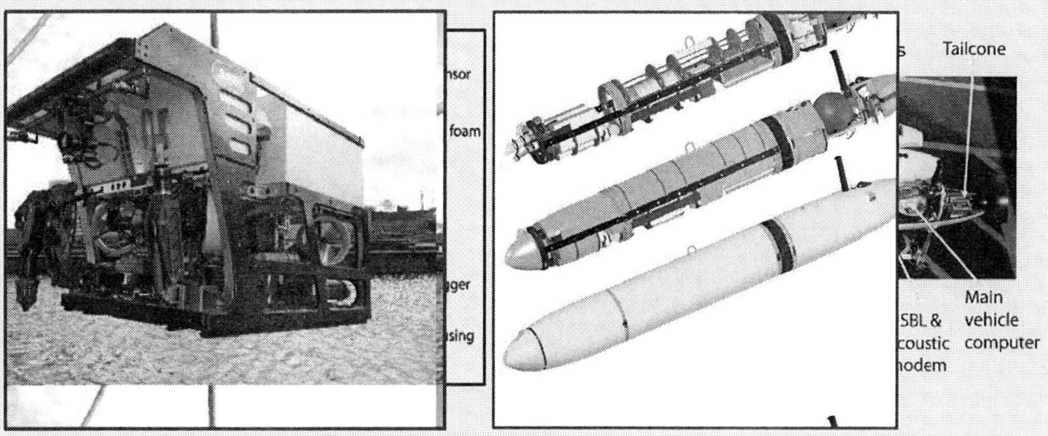

Higher density buoyancy material force use of larger buoyancy pieces; which increases encroachment into complex negative space between other components. More complex shapes = higher cost

Higher demand for lower density (higher grade) buoyancy materials

TRELLEBORG

Challenges in AUV Buoyancy

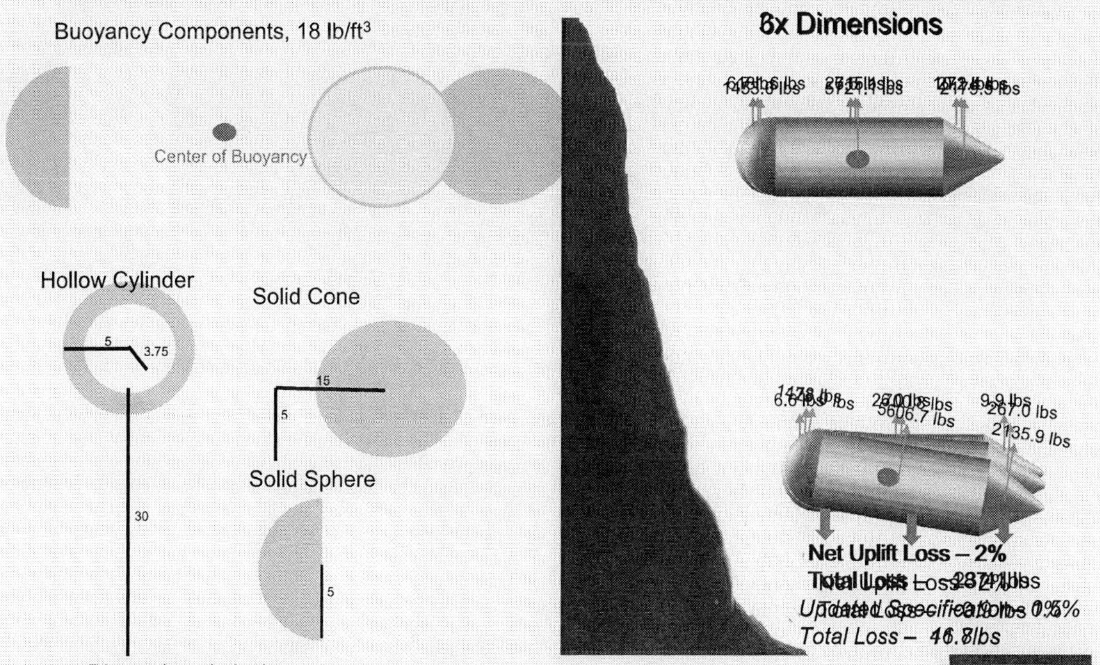

TRELLEBORG

Values are examples only; not representative of real specifications

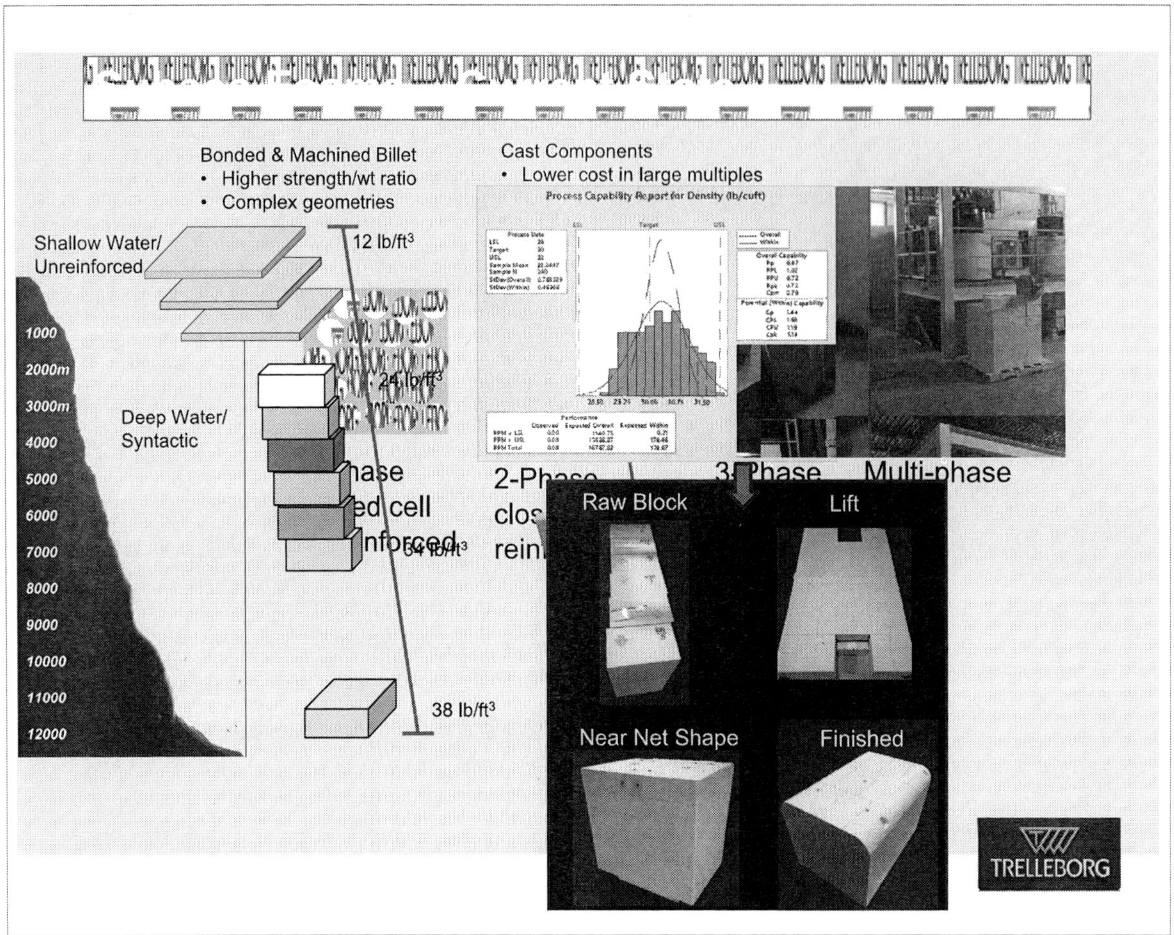
Bonded & Machined Billet
• Higher strength/wt ratio
• Complex geometries
Cast Components
• Lower cost in large multiples
Process Capability Report for Density (lb/cuft)
Shallow Water/
Unreinforced
Deep Water/
Syntactic
12 lb/ft³
24 lb/ft³
34 lb/ft³
38 lb/ft³
1000
2000m
3000m
4000
5000
6000
7000
8000
9000
10000
11000
12000
2-Phase
3-Phase
Multi-phase
Raw Block
Lift
Near Net Shape
Finished
TRELLEBORG

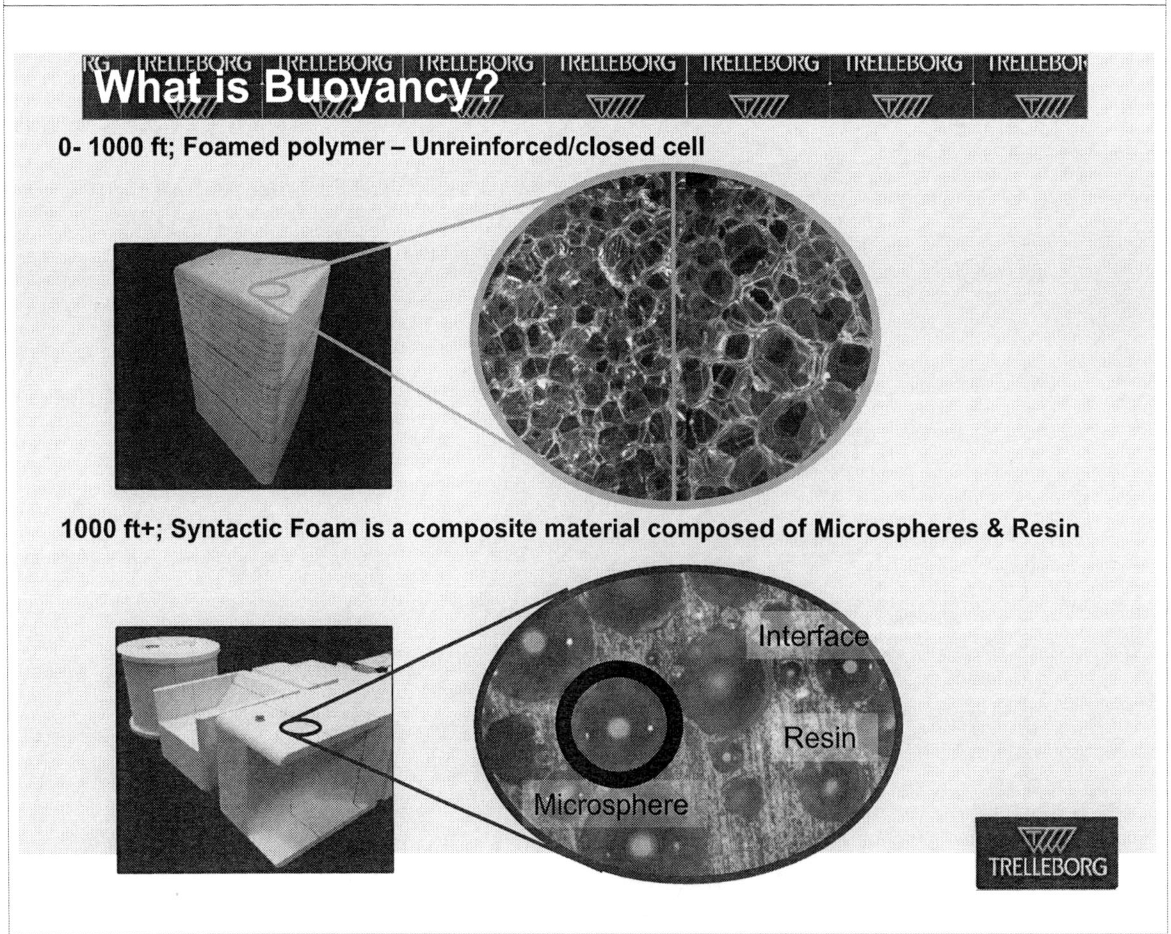
What is Buoyancy?
0- 1000 ft; Foamed polymer – Unreinforced/closed cell
1000 ft+; Syntactic Foam is a composite material composed of Microspheres & Resin
Interface
Resin
Microsphere
TRELLEBORG

Syntactic vs Shallow Water

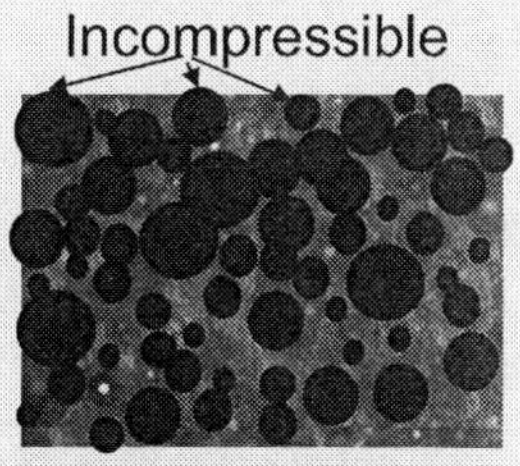

Values are examples only

Syntactic – Deep Water

- Waterpick up
  - Microsphere & Interfacial Failure (0 - 1 % weight gain)
- Bulk Modulus
  - Elastic compression of matrix at depth (0.5% vol loss)

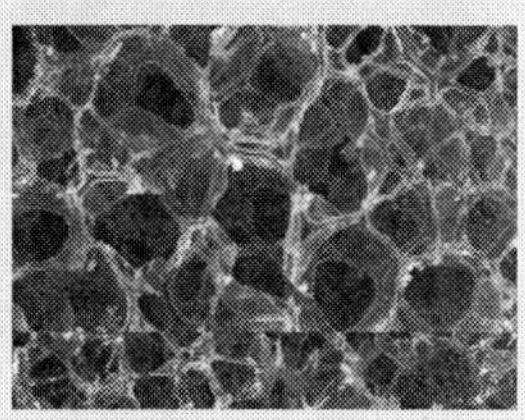

Closed Cell/ Unreinforced – Shallow Water

- Waterpick up Surface
  - Opencells (1% wt gain)
- Waterpick up Cell Failure
  - Cellwall rupture (1% wt gain)
- Bulk Modulus
  - Elastic compression (1% vol loss?)
- Polymer Creep
  - Plastic compression (1-5% vol loss)

7 Values are examples only; not representative of real specifications

TRELLEBORG

Deep Water Syntactic & Microspheres

What is a Microsphere?

- Silica Powder
- Spherical Shape
- Hollow
- Micro particle
  - Dust Mites ~ 200 micron
  - Hollow Glass Microspheres ~

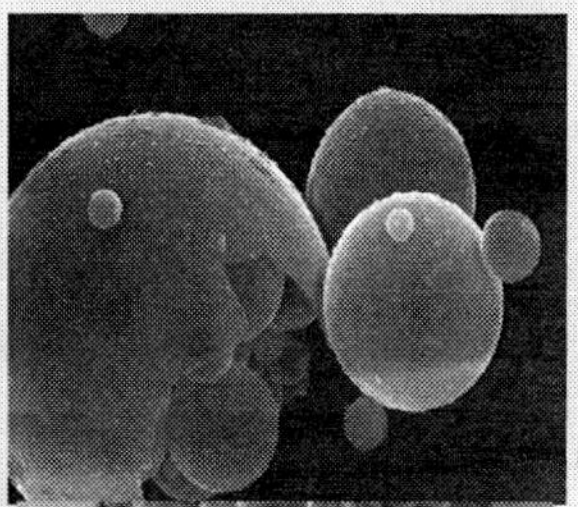

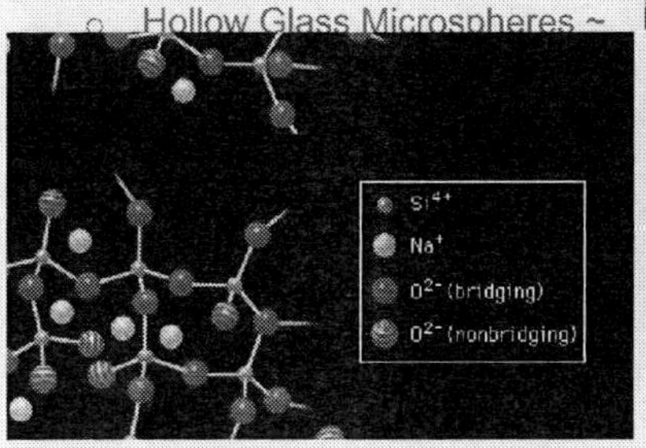

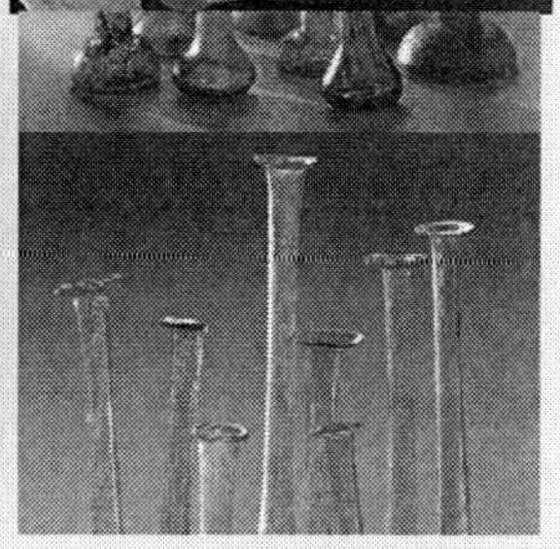

Hydrolytic Resistance (Resistance to Water Attack)

30% Sodium
Class V
(the worst)

<2% Sodium
Class I
(the best)

TRELLEBORG

Causes of Syntactic Foam Geriatric Aging

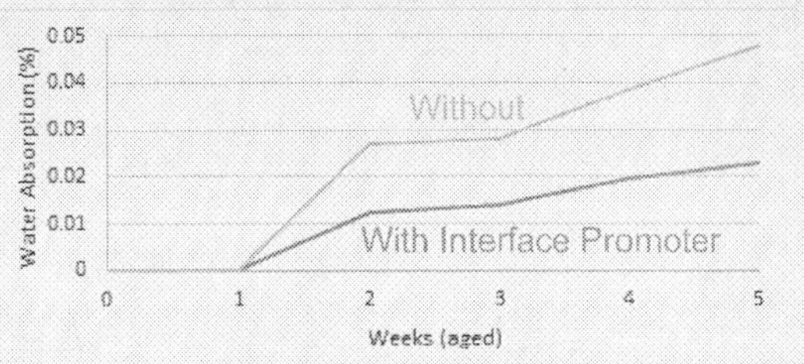

Tensile Properties

	W/O Interface Promoter	W/ Interface Promoter
Peak Stress (Mpa)	11.1	6.1
Strain at Peak (%)	1.8	2
Ext. at Break (%)	72	110

**Tensile properties in Syntactic Foam are an interface dominated failure

Microsphere Strength

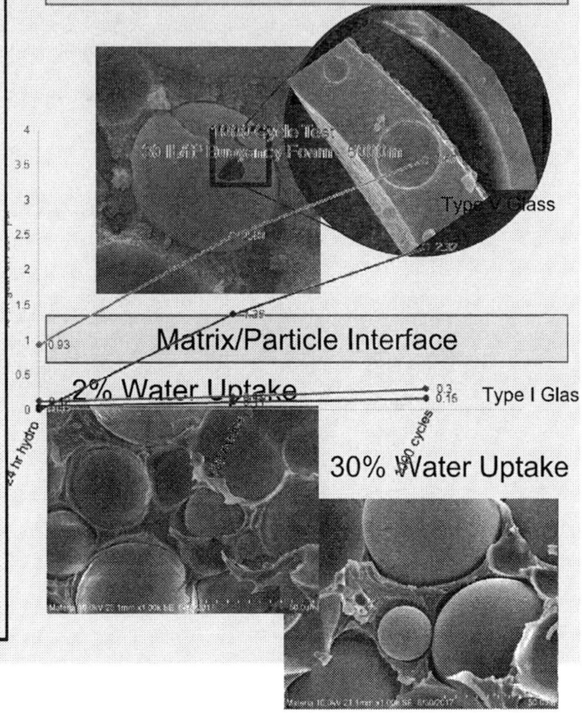

Shallow Water Materials (Unreinforced Foams)

Where do these materials come from?

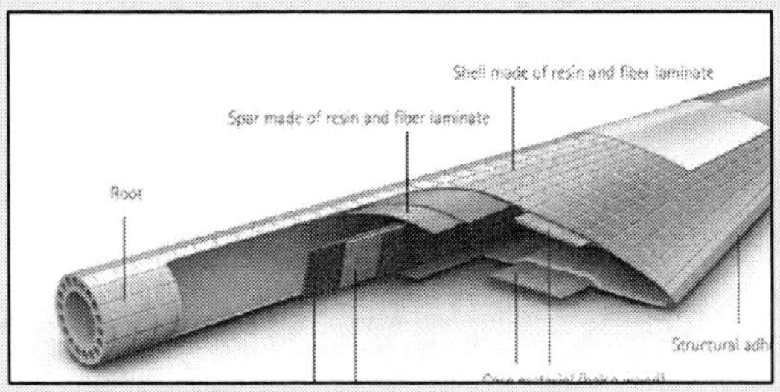

Surface adsorption 1 – 2 kg/m² at vacuum pressure ~20 psi

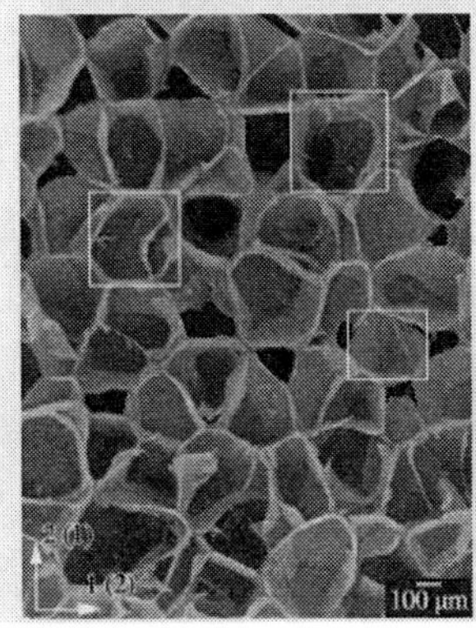

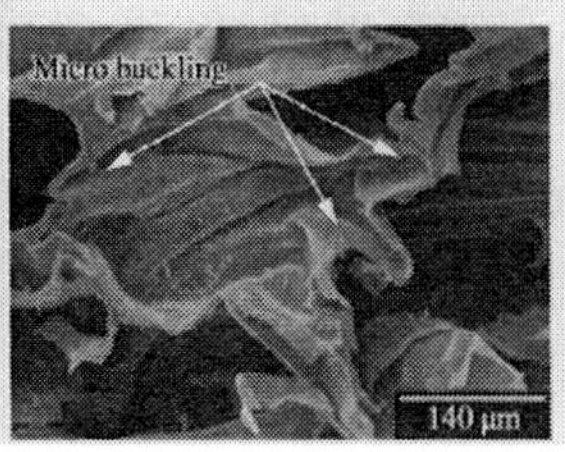

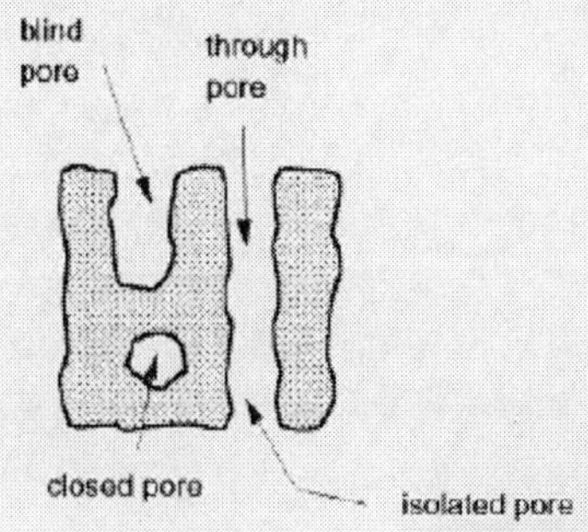

10

How Do They React at Depth

Unreinforced Shallow Water

Syntactic Deep Water

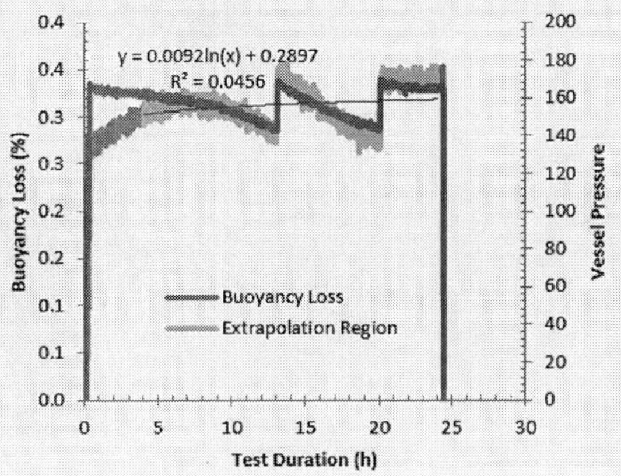

11 TRELLEBORG GROUP

TRELLEBORG

Hydrostatic Loading of Buoyancy Materials

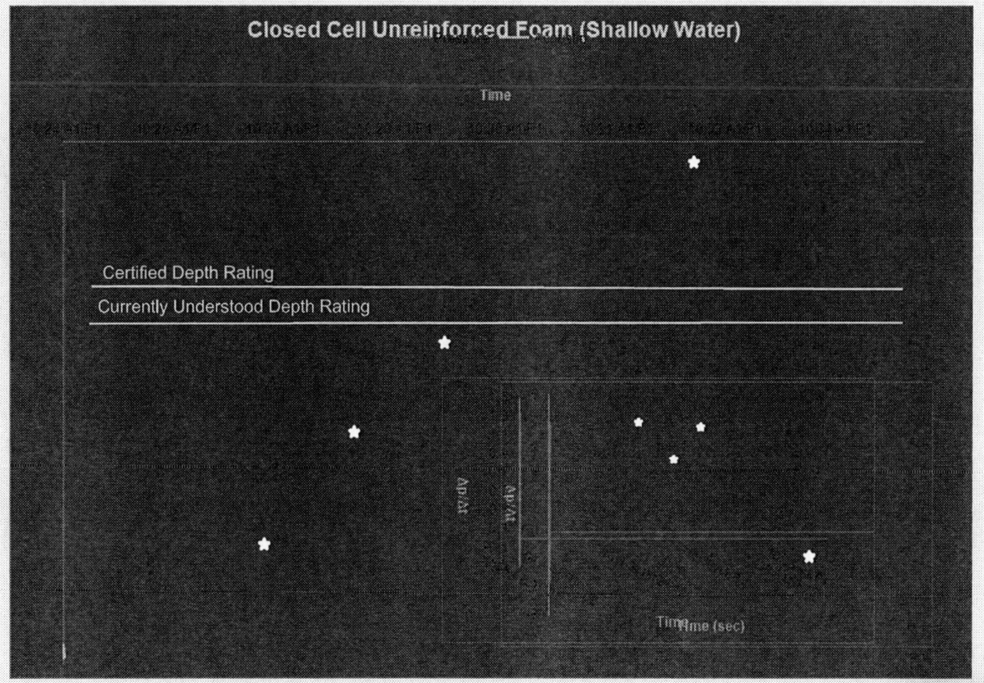

12 TRELLEBORG GROUP

TRELLEBORG

FLow Density Thermoplastic Feedstock

Convert Components into Structurally Integral Buoyancy
- o Fairing Components
- o Internal Ribbing, Plates
- o External Fins & Skid Components

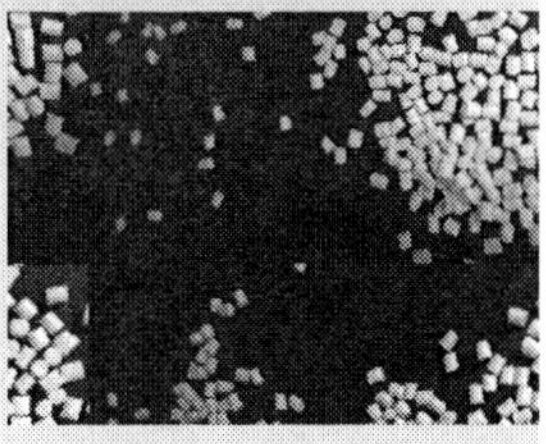

Molded/Machined Billets

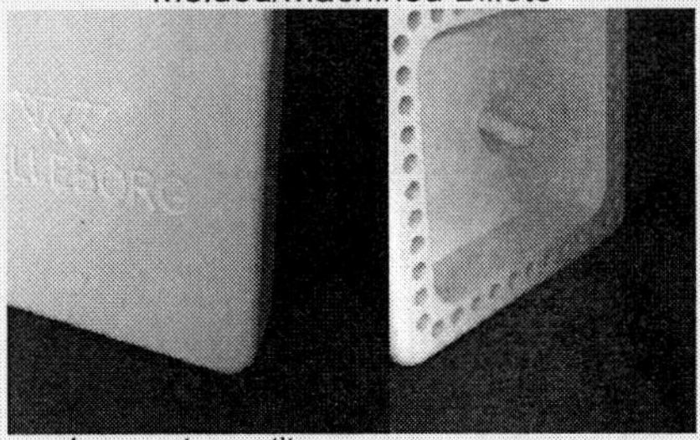

- Low part quantity
- Higher part cost (comparatively)
- Large size parts

Injection Molded Pieces

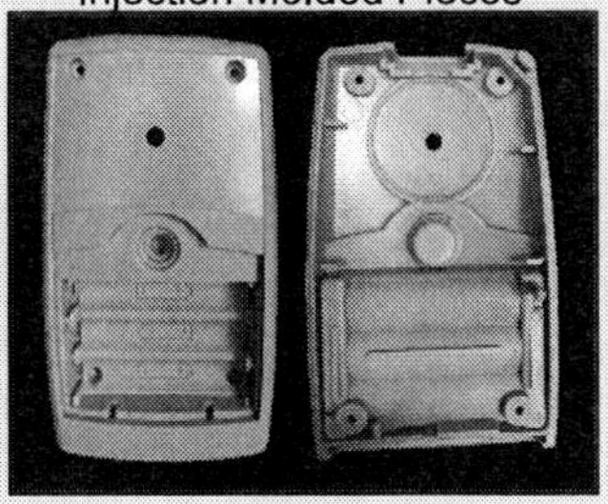

- High part quantity (100's +)
- Low part cost
- Large upfront tool cost
- Small/medium part sizes

3D Printing

- Prototypes
- Intricate one-offs/ small quantity
- Printed into ultra tight spaces
- *Skys-the-limit*

TRELLEBORG

TRELLEBORG

Outline

- **Overview**
- **ASV C-Worker Class ROV-Capable Designs**
- **Applications of the ASV Technology in Offshore Industry**
- **ASV Piloting, Navigation Aids & User Interface**
- **Specific Project Examples**
- **Conclusions**

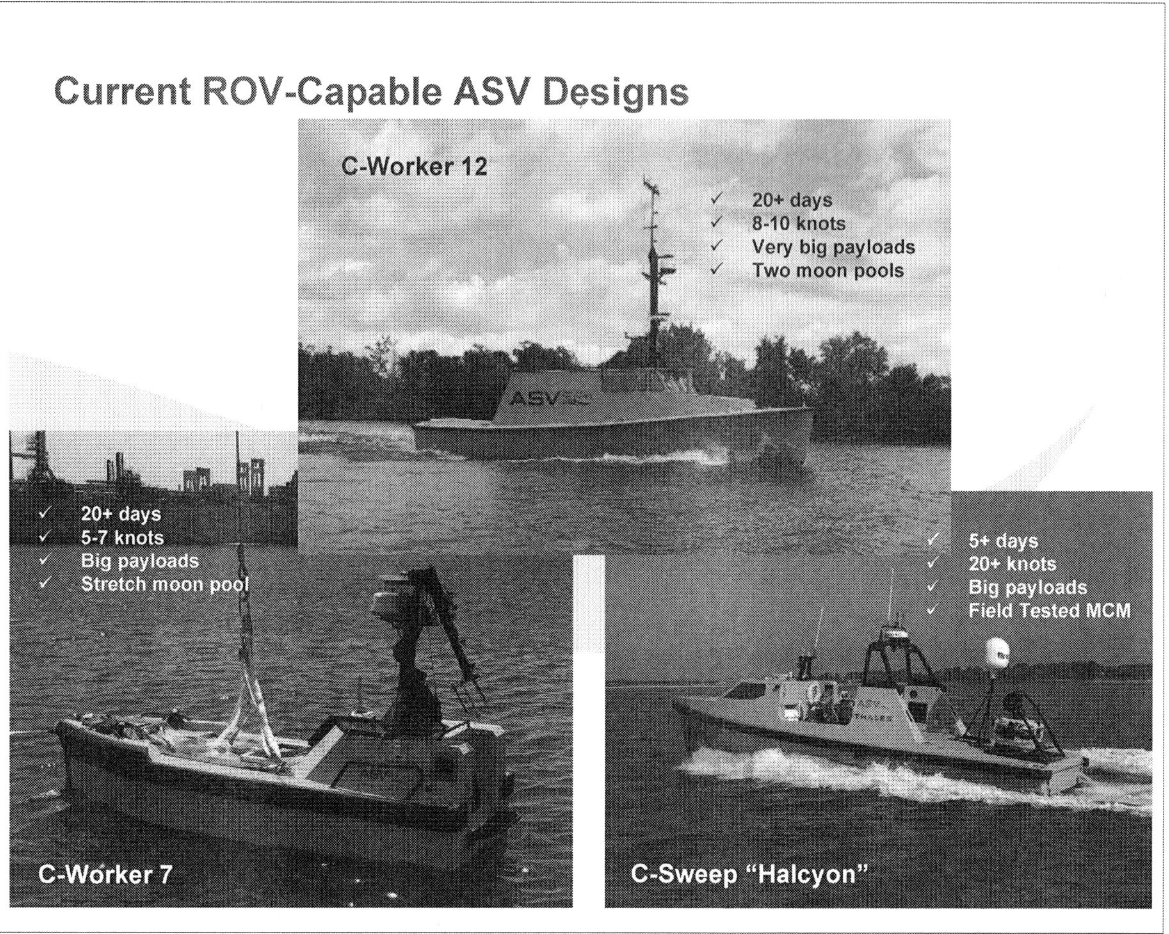

Applications to date:

- **Hydrography** – NOAA and other seabed charting operations
- **Cable Route Surveys** - Alaska
- **Marine construction survey support -** (USBL, TDM/ROV tracking, LBL array setup, As-Built Surveys)
- **Environmental** – Passive acoustics for mammals & fish, seep/leak detection, water quality, LiDAR, oil dispersant and skimmer systems
- **Data Conduit** – Low cost upload from seabed data loggers; positioning & data for AUVs and buoyancy gliders
- **Security** – Harbor and offshore security, 24/7 unmanned monitoring, cameras, IR, other
- **Metocean** – ADCPs, Loop Current monitoring
- **Seismic survey support** – Unmanned Airgun Arrays, USBL, CTD, PAM, ADCP
- **Military** – Targets, MCM, AUVs/ROVs, surveillance, scouting
- **Towed Equipment** (Sidescan, Subbottom Profiler, Mags, UXO, etc.)
- **Conversions** – RIBs, FRCs, Patrol Craft, Crew Boats, Others

Operational Video – Target & MCM Operations

C-Worker with Oceaneering SeaMaxx Obs Class ROV

C-Worker with Oceaneering SeaMaxx Obs Class ROV

ASView – Control System & User Interface

Top down displaying;
Full S57 ENC's
AIS targets from receiver on USV
Vessel track
Vessel mission plans
4 quadrant HD Cameras
Scanning IR Camera
3rd Party Software Windows, (Hypack, etc)

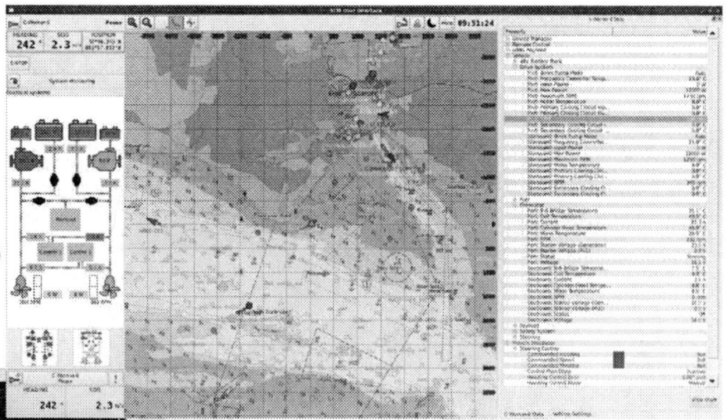

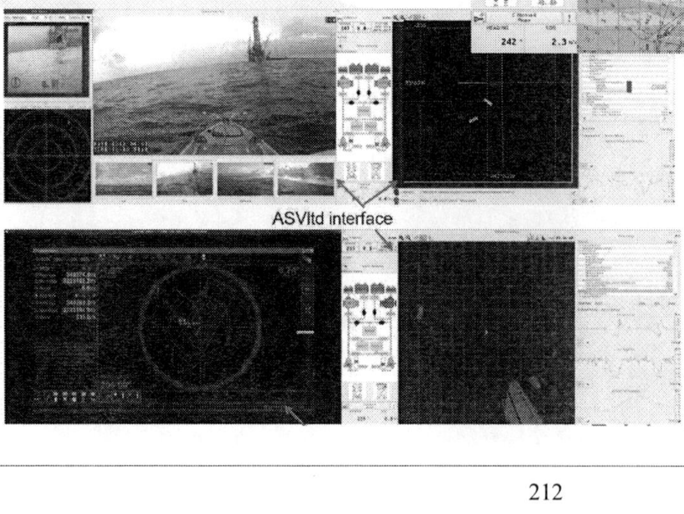

Project Examples - Egypt Pipe Lay Project

The C-Worker 7 autonomous surface vessel was deployed to the Mediterranean to support a pipelay operation for Subsea 7. The C-Worker 7 was used for pre-lay survey, touchdown monitoring and As-Laid Survey.

Benefits: Big Cost Reduction, Wider Weather Window, Decreased HSE Risk

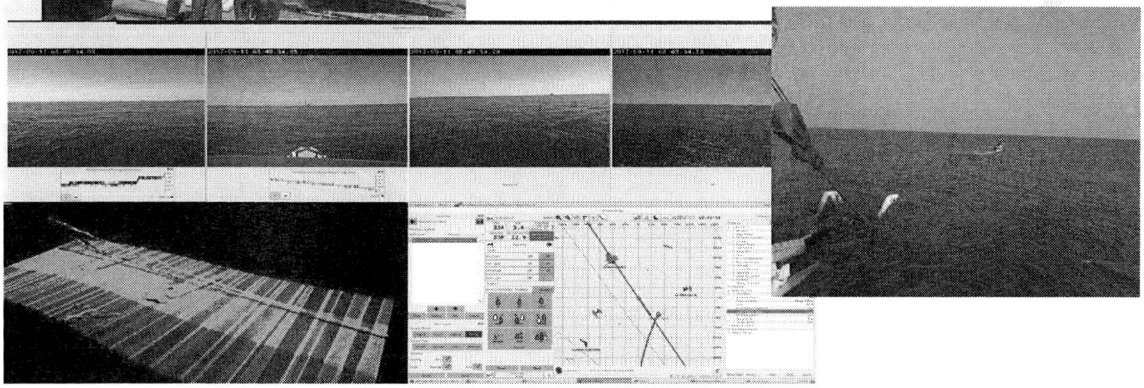

Project Examples – Subsea Tieback Project

LBL Array Calibration and Cost Savings

- The decision was made to use the ASV to complete the box in's thus allowing the vessel to continue with other scheduled tasks. During this time the survey team performed a test to evaluate the possibility of the ASV collecting baselines and performing the calibration. The test was successful and they proceeded to calibrate the array with the ASV thus freeing the ROV to continue with other tasks.
- Total estimated vessel savings time for the LBL installation (bottom example is Stampede):

Array	# Transponders	#Boxin	Time(3h/Box)	Calibration	Total (Hrs)
Flowline	9	3	9	5	14
In Line	25	5	15	15	30
Umbilical	4	2	6	1	7
				Total Savings	51

Grant Candies Vessel Time Savings to SS7 (based on prior 3hr/Boxin estimates by SS7):			
No. Boxins	Avg. ASV Hrs per Boxin	Time Saved Per Boxin	Total Time Savings (Hrs)
25	**0.65**	**3**	**75**

Project Examples – C-Worker 8 & AUV Support

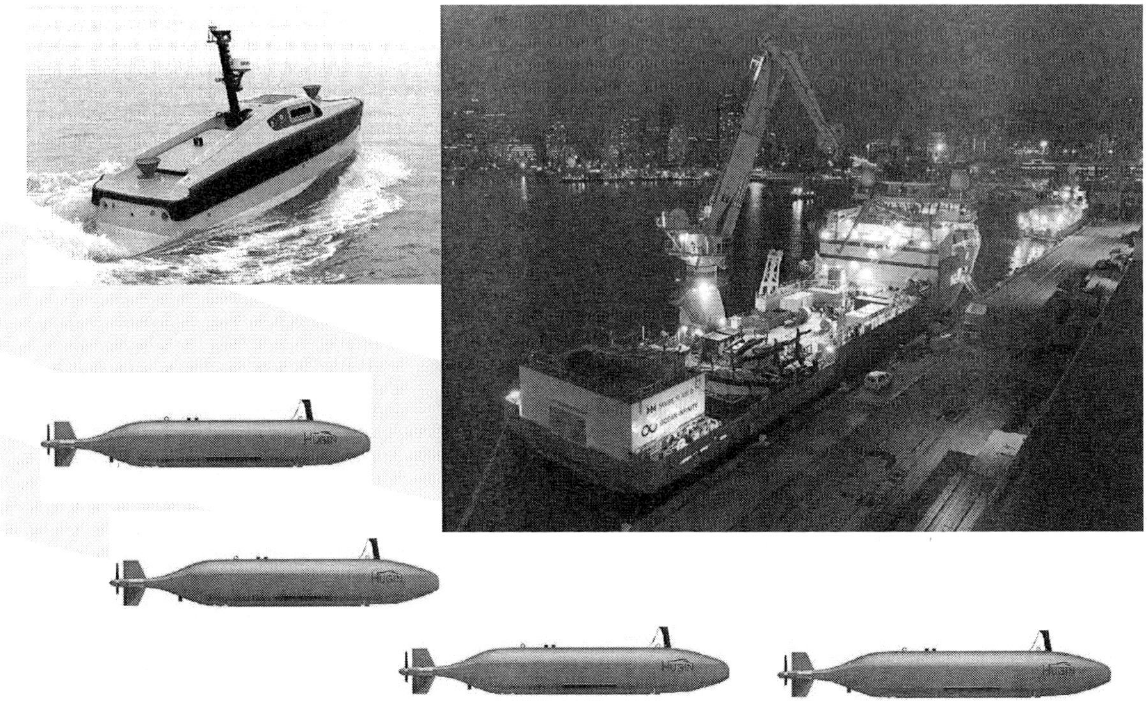

Project Examples – C-Worker 8 & AUV Support

Project Examples – Seismic Airgun Source Project

A Novel Small Source Platform: Rapid Autonomous Marine 4D (RAM4D)

- Unmanned & autonomous seismic acquisition
  - Off-the-shelf autonomous/unmanned diesel vessel
  - Small airgun source (size/compressor limited)
    - Notionally 250 in^3
  - Planned 10-15 day endurance
  - Remotely supervised, autonomous operation
  - Targeting ocean-bottom cables, nodes, DAS

Photo courtesy of PGS (www.pgs.com)

Vessel provided by ASV Global Inc., Float by Ulmatec Baro

- Benefits of RAM4D:
  - Lower source costs ($5k-15k vs $50k per day)
  - Greatly reduced HSE exposure
  - New market segment
  - Flexible and cheap acquisition mobilization

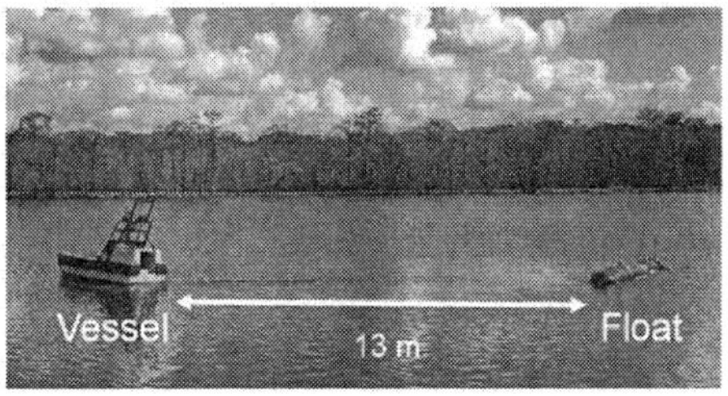

Target:
10x reduction in vessel size, source size and cost

Shell

Project Examples – Seismic Airgun Source Project

Project Examples - Loop Current Profiling - GoM

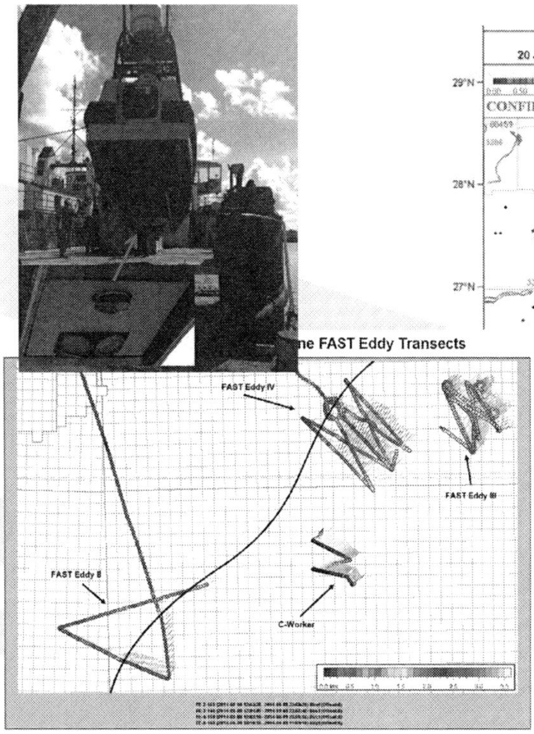

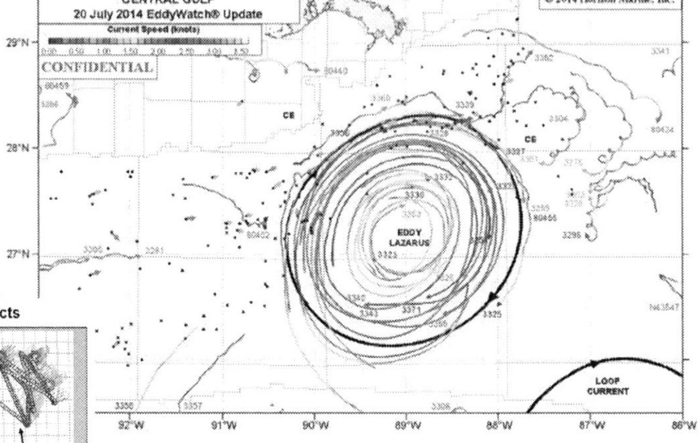

GoM work with Horizon Marine monitoring Loop Current

Project Examples – Hydrographic Survey Projects

- **C-Worker 5 worked alongside the Qualifier Q105 survey vessel for 36 days duration**
- The C-Worker 5™ **completed 2275 nautical miles** of survey lines, operating as a force multiplier running parallel survey lines to the Q105.
- Total **ASV work was 44% of the project total**, above the projected 30-35%
- Both vessels had identical Reson 7101 multibeams and simultaneously towed the same Edgetech 4200 side-scan sonars.
- **In 2017 Total ASV Alaska work was 53% of the project total for NOAA.**

Project Examples – Wind Farm Survey Projects

- Confirmed ability to acquire high quality wind farm site surveys
- Permitted wind farm area surveyed
- 3 wreck surveys
- Similar production rate to conventional methods at lower cost
- Very clean data – provides benefit to customers for reduced processing requirements

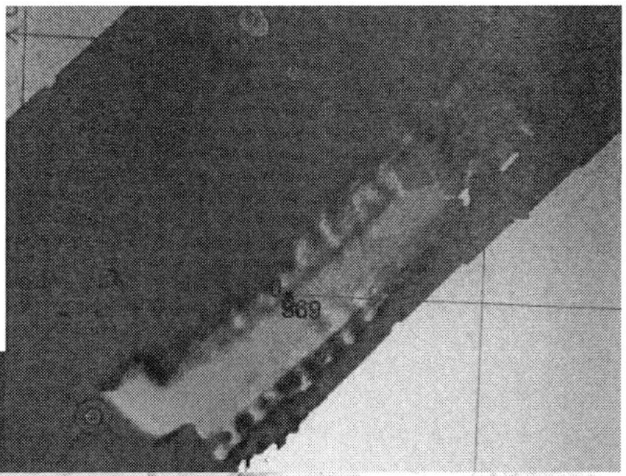

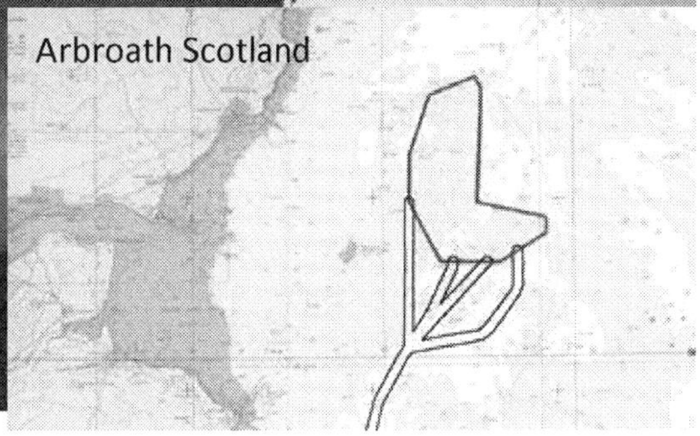

ROV-Capable ASVs - Summary

Benefits

- Easily mobilized by sea or air
- Decreases man hours on the water, improves HSE risks
- Robust designs, heavy duty
- Flexible payload for multiple parallel applications
- Each system includes LARS gear
- Significant cost savings over surface ship operations

Conversions

- Multiple units being fitted

Credits and Acknowledgements

- SeaTrepid
- Ocean Infinity
- Shell
- Subsea 7
- Horizon Marine
- NOAA
- Terrasond Ltd.
- Thales
- Oceaneering

Welcome To Focal Technologies

Reducing ROV Operating Costs and Enhancing Reliability with Monitoring Technology

Ian MacKay, P.Eng.

ROV Health Monitoring Systems

Reducing Downtime by Live Health Monitoring in Three Areas:

- **Optical multiplexer system for TMS and ROV telemetry/video/control**
- **High Voltage/Current Slip Rings - critical rotary components.**
- **Optical cable system from surface to vehicle**

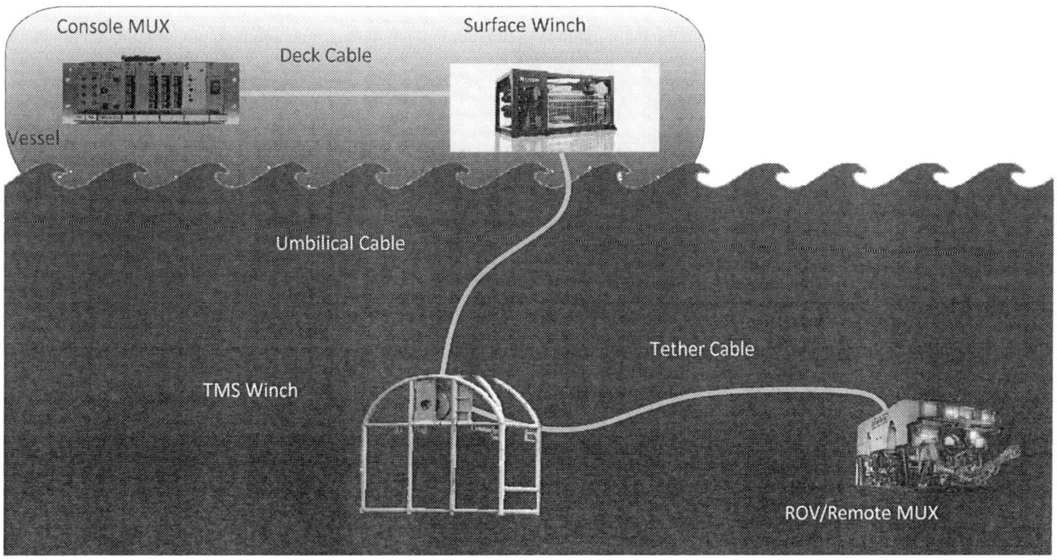

Optical Cabling System Health Monitoring

Sample WROV Optical Communication Path

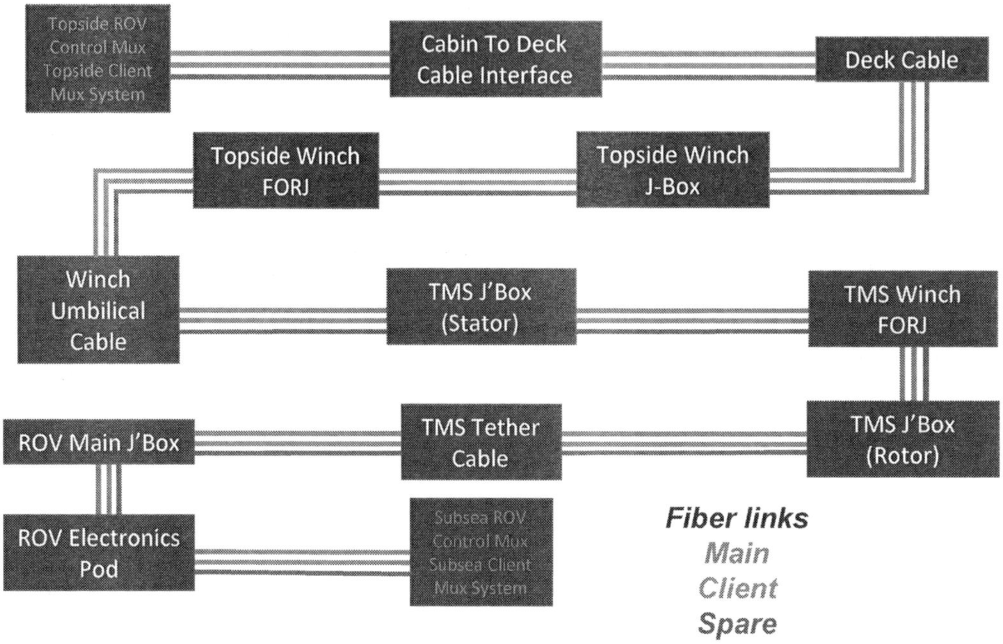

Optical Cabling System Health Monitoring

What happens when optical links fail…present day?

❖**Dig out the expensive Optical Time Domain Reflectometer (OTDR)**

- **Requires OTDR trained ROV technicians**
- **Offline system…need to break into optical channels**
- **Difficult to interpret results, many pitfalls**
- **Difficult to locate subsea faults when ROV deployed**
- **Fault diagnosis is slow… more downtime!**

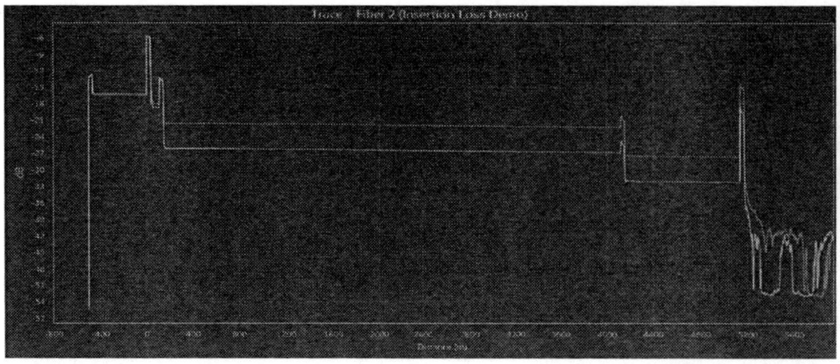

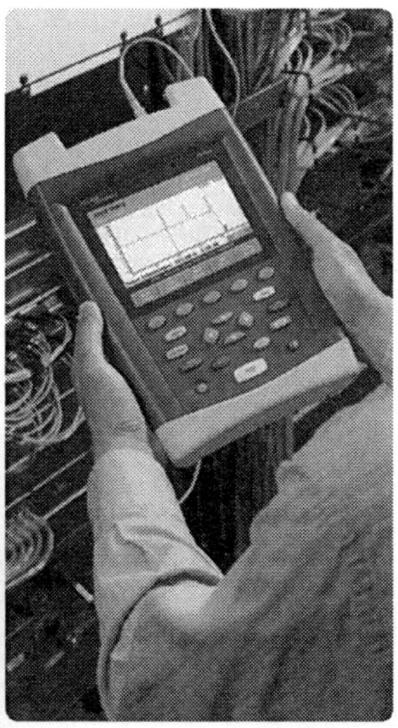

Optical Cabling System Health Monitoring

New Capabilities with Focal Optical Monitoring System (OMS)

- **Real time monitoring** of up to 4 optical links at the same time
- **Operates inline with live telemetry data** and on spare fibers
- **Simple and easy to use**...no special training required, HMI style GUI, automatically runs on power up
- **User configurable alarms and warnings** in GUI and panel LEDs
- **Up to 32 fibers** supported for subsea control applications
- **Central diagnostic hub** integrates with Focal mux diagnostics, slip ring sensors, and 3rd party data, e.g. NMEA-0183, with local data logging
- **Network interface available** for remote access and remote data transfer, e.g. cloud

Optical Cabling System Health Monitoring

User configurable GUI: OMS, Multiplexer, Slip Ring Diagnostics

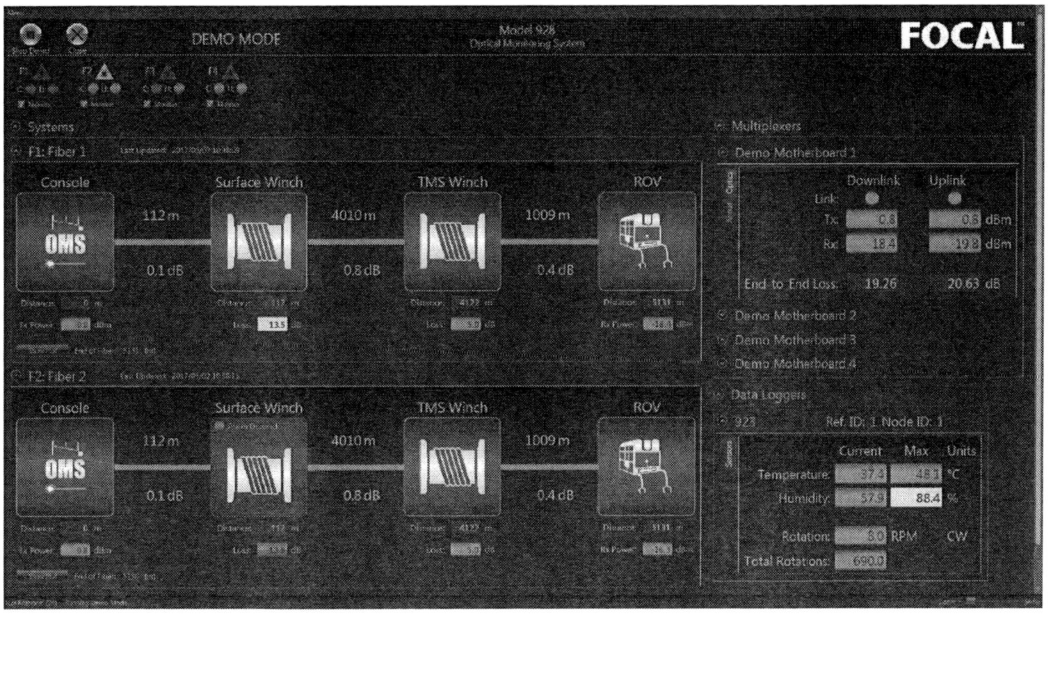

Optical Cabling System Health Monitoring

Traditional OTDR traces with saved reference

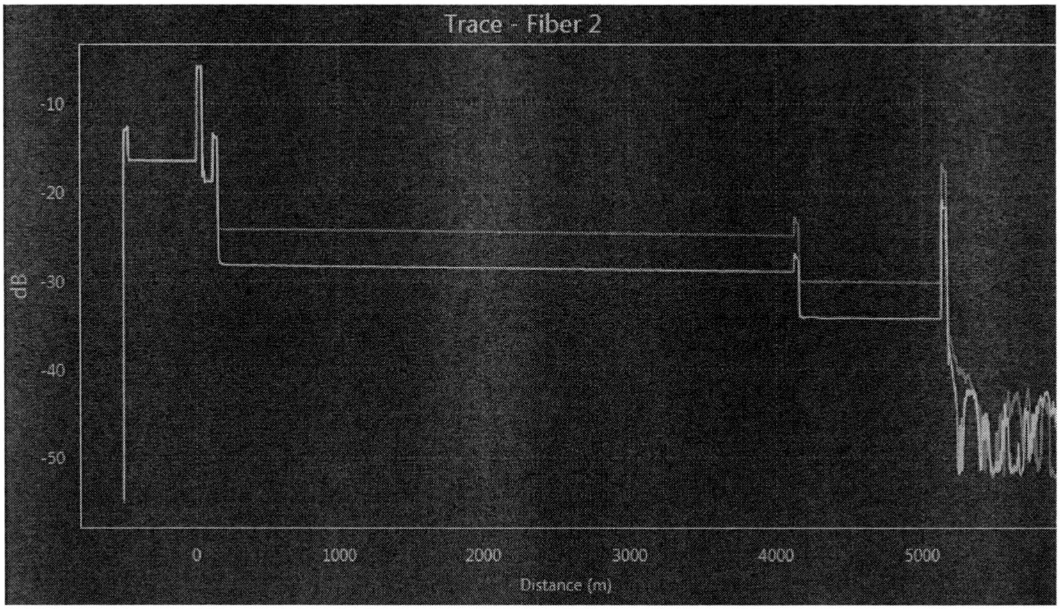

Optical Cabling System Health Monitoring

Sample WROV Optical Communication Path – OMS Added To Circuit

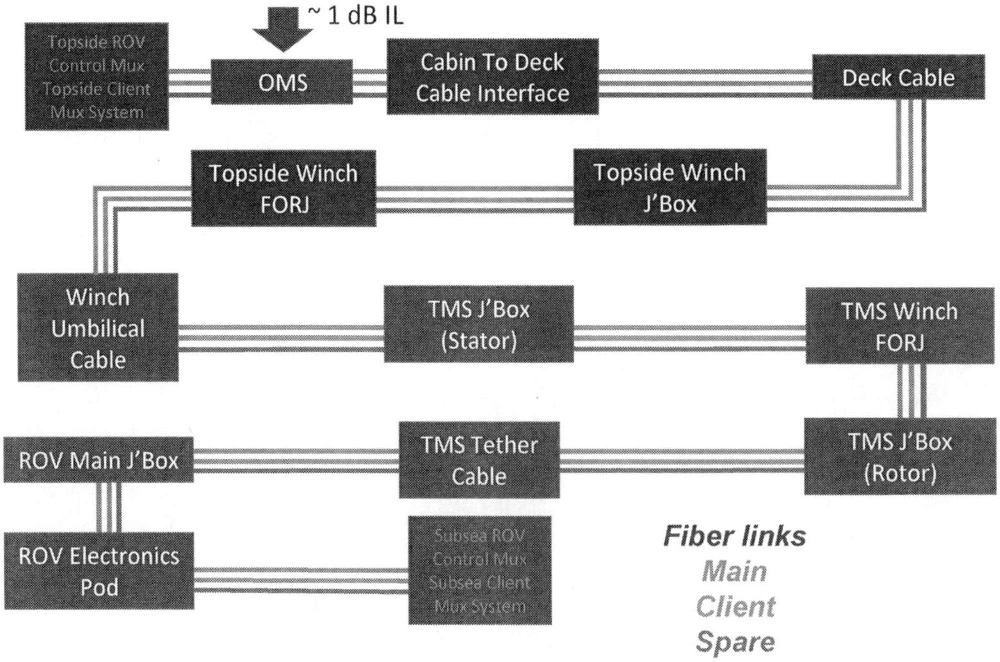

Optical Cabling System Health Monitoring

User interface showing system operational.

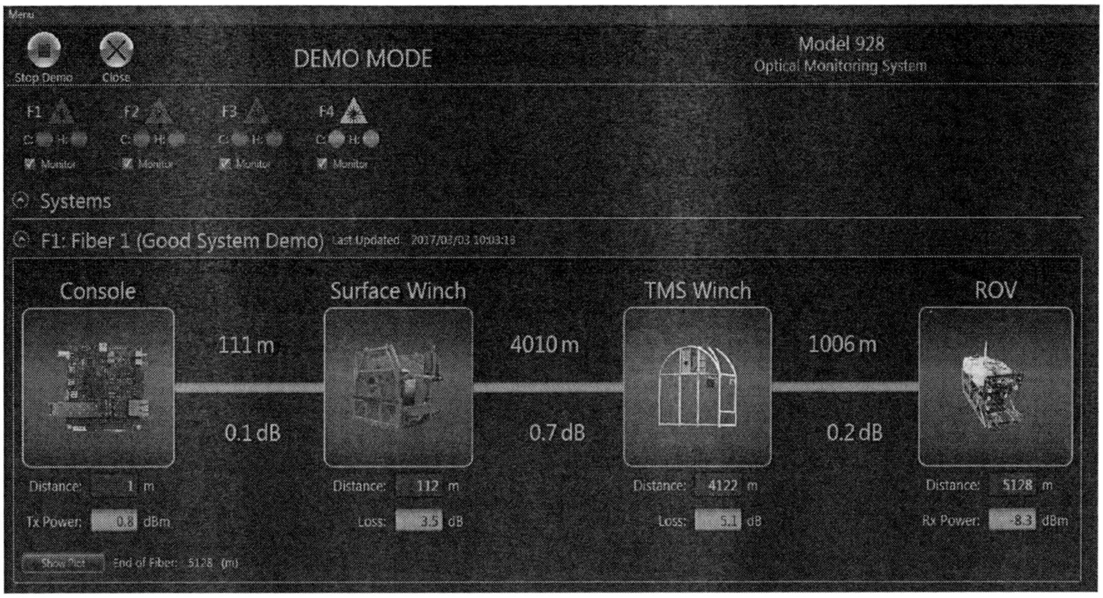

Optical Cabling System Health Monitoring

User interface showing insertion loss failure on fiber 2 in surface winch.

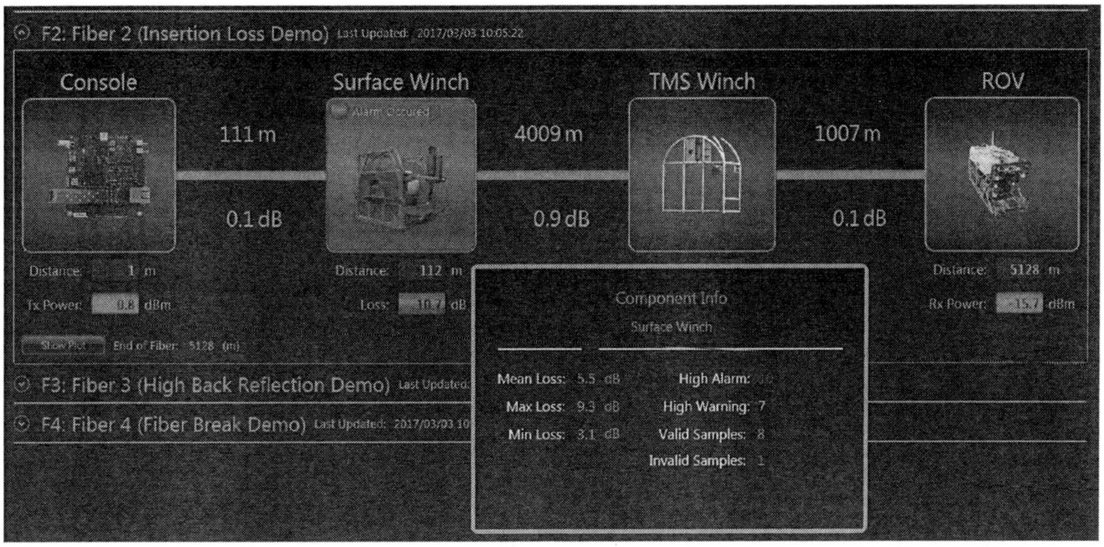

Optical Cabling System Health Monitoring

User interface showing open circuit on fiber 4 in tether.

Expected tether length is 1300m

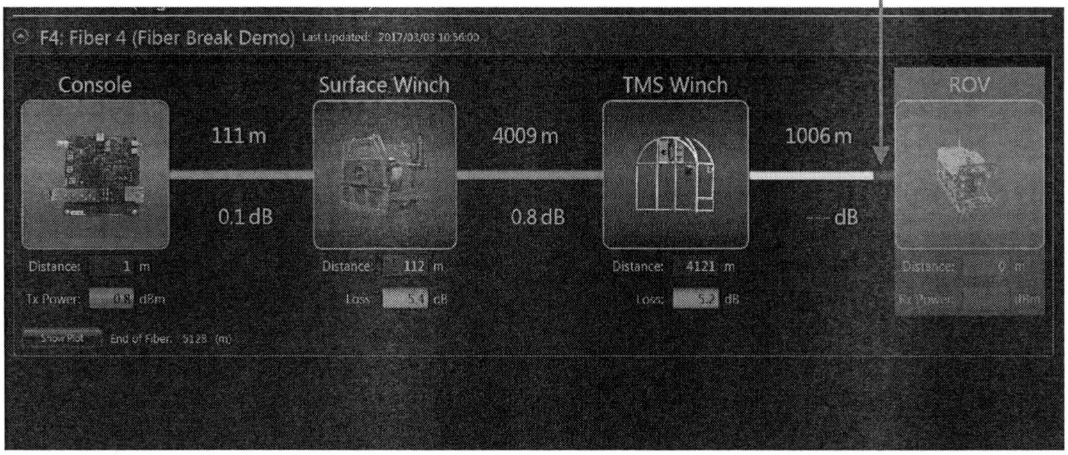

ROV Slip Ring Health Monitoring

- ❖ **Focal Slip Rings typically installed in Surface and TMS winches**
- ❖ **Rated up to 7.2kV at 20A for power lines and 600V at 7A for signal lines.**
- ❖ **Electrical slip ring usually combined with Fiber Optic Rotary Joint (FORJ)**
- ❖ **TMS slip ring usually immersed in oil and pressure compensated.**

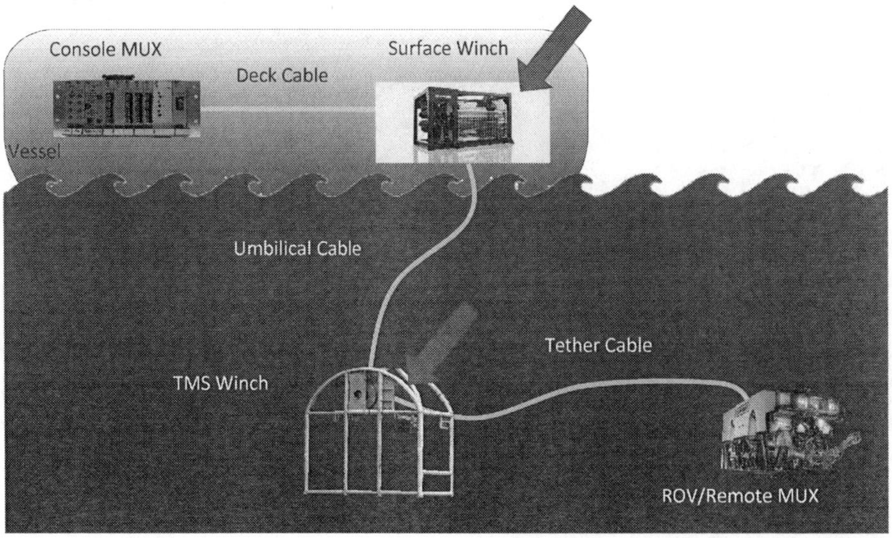

ROV Slip Ring Health Monitoring

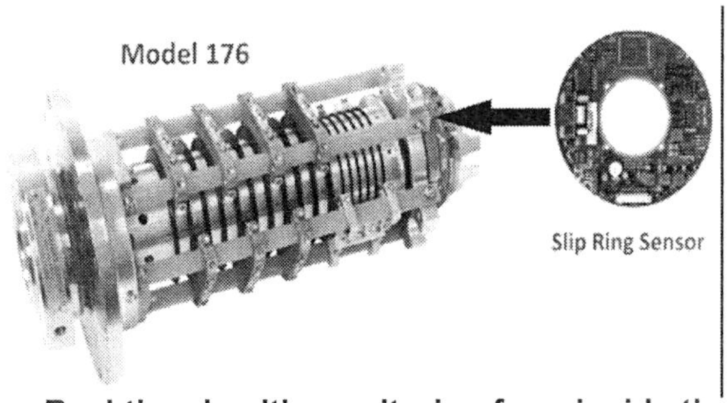

- Real-time health monitoring from inside the slip ring assembly
- Improves uptime and guides planned maintenance activities
- Designed for new configurations or retrofit into existing slip-rings with minimal impact
- Optional external sensors including water ingress, current, pressure, voltage, etc.
- Qualified for use in surface and subsea slip rings (up to 4,000m depth rated)
- "Black Box" onboard data storage, min. 5 years recording life

Sensors	
Temperature	Onboard -20 to +70 °C
	2 channel T/C, K type
Relative Humidity	5 to 95% RH
Accelerometer (3-Axis)	± 200 g per axis with shock event detection
Ambient Light	300 to 1000 nm
Turns Counter	Direction, speed, total turns (10 degree resolution typical)

Diagnostics Serial Link	
Signal Type	RS485
Speed	115.2 kbps
Protocol	Modbus RTU

ROV Slip Ring Health Monitoring

User Interface…all electrical slip ring operating parameters good

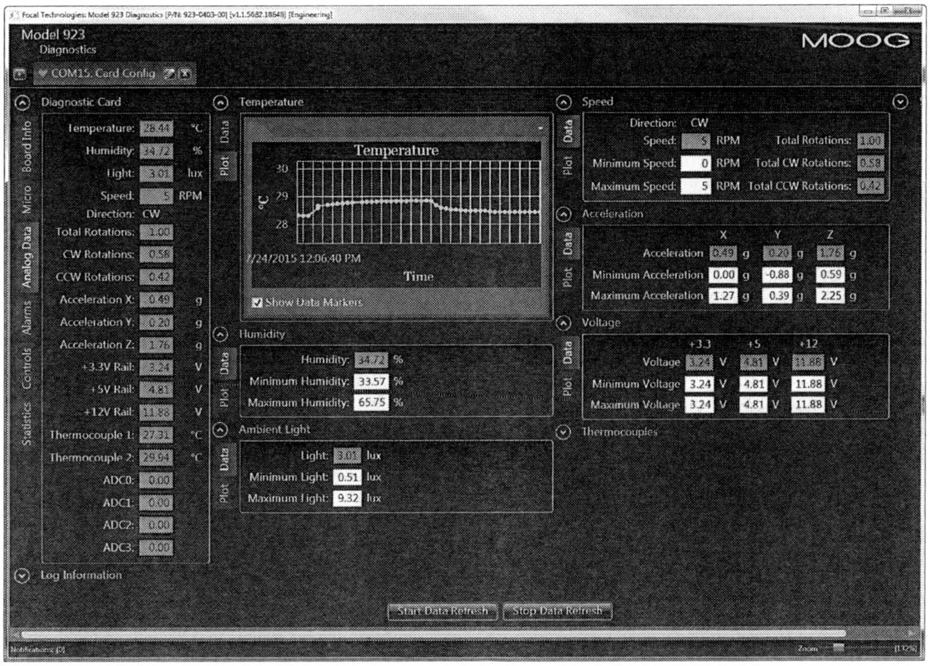

ROV Telemetry Health Monitoring

Model 907 Diagnostics

- **Health monitoring of topside and subsea communication data links**
- **Temperature, laser bias current, power-rail voltage, Tx/Rx optical power**
- **Remotely configure switch settings without opening bottle on newer cards (907E)**
- **Seamless integration into Focal health monitoring software**
- **Software Developer Kit available**

ROV Telemetry Health Monitoring

User Interface...serial data and optical links status

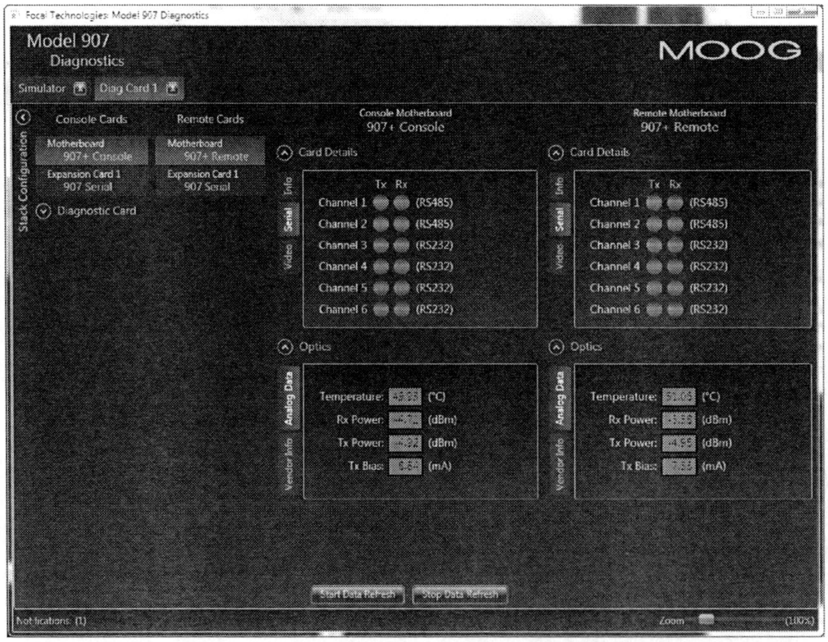

Health Monitoring System – Hub and Gateway

Comprehensive Condition Monitoring System:
Diagnostics + Optical System Monitoring + Slip Ring Sensors

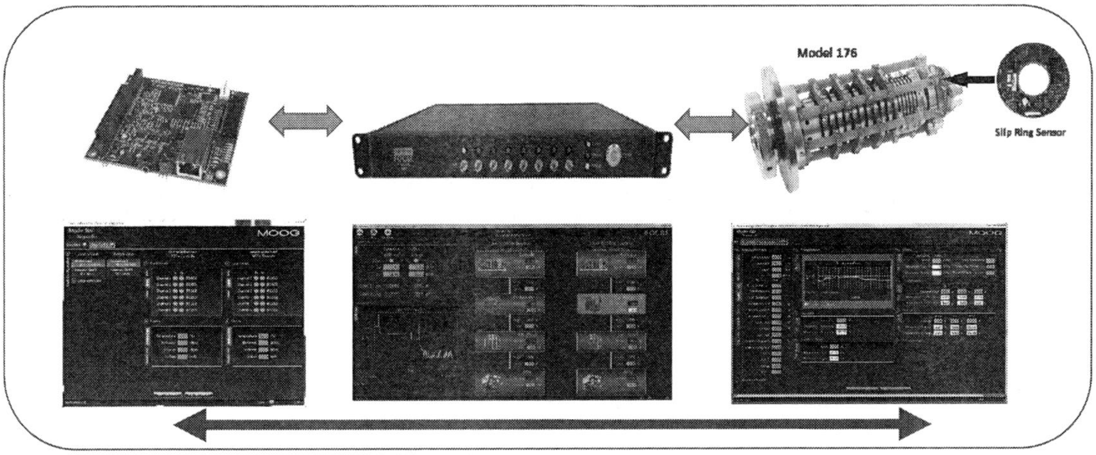

- ✓ *Linked hardware to collect and manage data*
- ✓ *One easy-to-use software platform and API*
- ✓ *Remote access and cloud-based data*

Health Monitoring System – IoT and the Cloud

OMS acts as hub for collection of all system diagnostics and gateway for IoT access to cloud services.

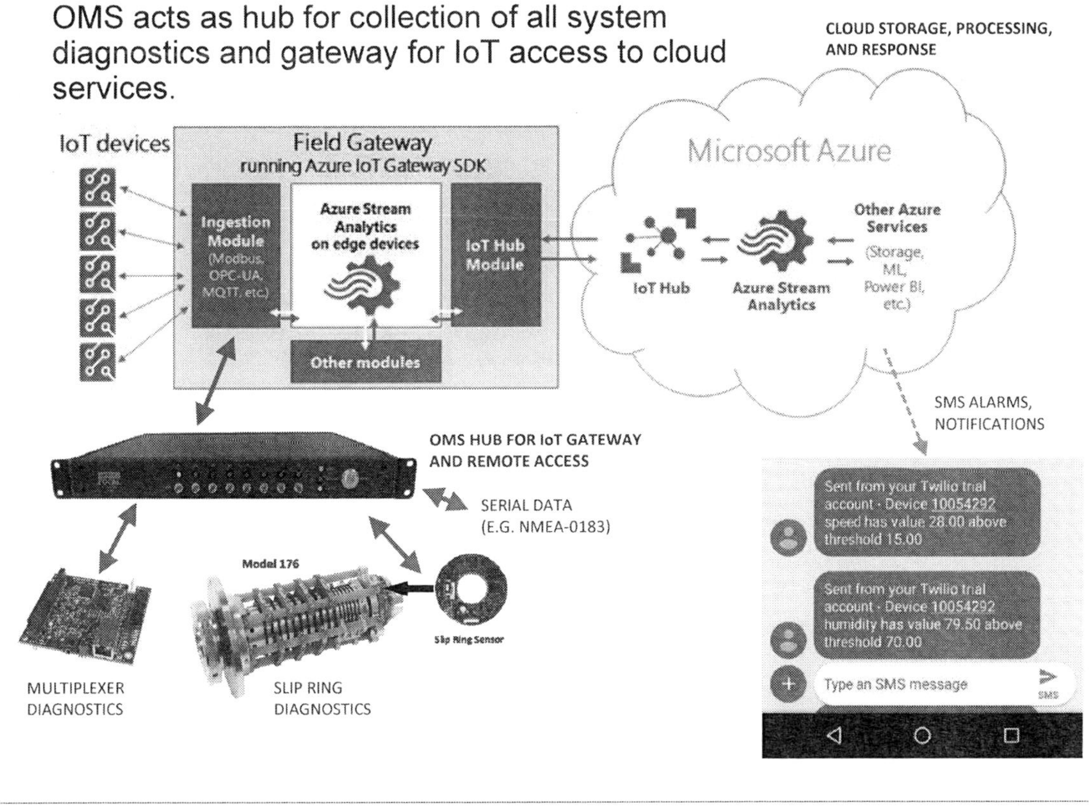

MOOG

Need more information?

Web: www.moog.com

Email: focal@moog.com

Tel: (902) 468-2263

Ian MacKay, P.Eng.
Electronics Engineering Manager
Focal Technologies Corp.
Moog Industrial Group
77 Frazee Ave., Dartmouth
Nova Scotia, Canada B3B 1Z4
imackay@moog.com

Contact your local MacArtney office

MacArtney
UNDERWATER TECHNOLOGY

MATE underwater robotics

SeaMATE ROV Kits Progression of Learning

The MATE Center's kits and competition provide steps for students to increase their knowledge and skills.

www.marinetech.org
MARINE ADVANCED TECHNOLOGY EDUCATION CENTER

1

MATE underwater robotics

History

The MATE Center started with focus on community colleges and upper level ROVs.

Competition was (and still is) open to any design. Safety and specification guidelines, but no rules on platforms.

www.marinetech.org

2

MATE underwater robotics

History

High schools also competed in the early competitions.

Middle schools wanted to join the competition.

Some upper elementary schools wanted to join the competition.

So MATE created educator workshops to start teaching how to build ROVs.

www.marinetech.org

3

MATE underwater robotics

History

MATE found that some educators were fine finding all the parts and pieces on their own, and modifying a build. But other teachers had troubles sourcing all the components.

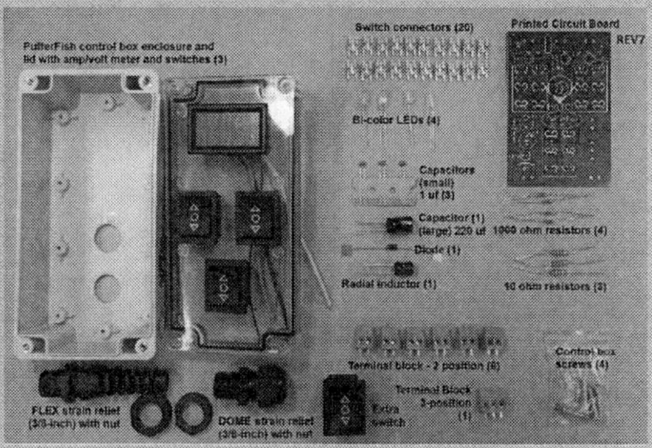

www.marinetech.org

4

MATE underwater robotics

AngelFish ROV kit

Started with a simple switch box controller called the AngelFish.

Simple switch box controller

Three DPDT (double pole, double throw) switches controlling three motors.

www.marinetech.org

5

MATE underwater robotics

AngelFish Lessons

Great for beginners with no robotics experience

- How a DPDT switch works
- How electricity flows through a circuit
- Soldering wires together

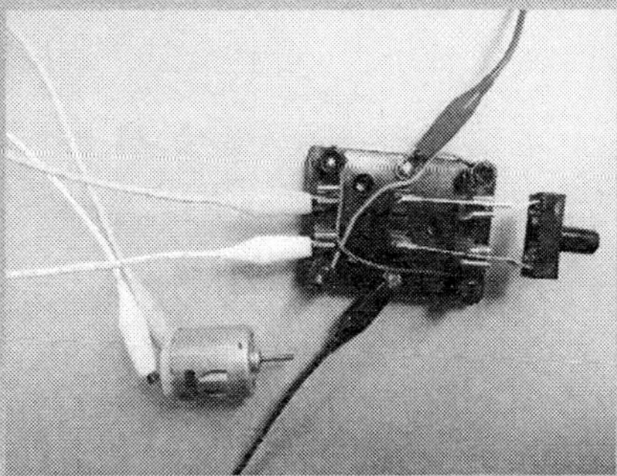

www.marinetech.org

6

MATE underwater robotics

AngelFish ROV kit

Easy construction.

No experience needed.

Basic easy to follow lessons.

Simple tools only.

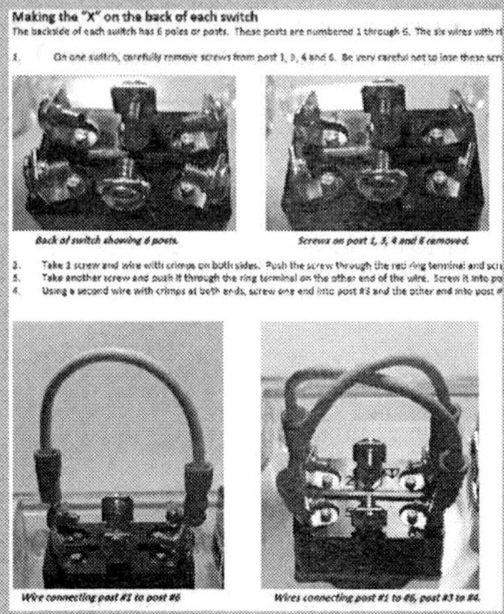
Making the "X" on the back of each switch

The backside of each switch has 6 poles or posts. These posts are numbered 1 through 6. The six wires with r

1. On one switch, carefully remove screws from post 1, 3, 4 and 6. Be very careful not to lose these scr

Back of switch showing 6 posts.

Screws on post 1, 3, 4 and 6 removed.

2. Take 1 screw and wire with crimps on both sides. Push the screw through the red ring terminal and scr
3. Take another screw and push it through the ring terminal on the other end of the wire. Screw it into po
4. Using a second wire with crimps at both ends, screw one end into post #3 and the other end into post #

Wire connecting post #1 to post #6

Wires connecting post #1 to #6, post #3 to #4.

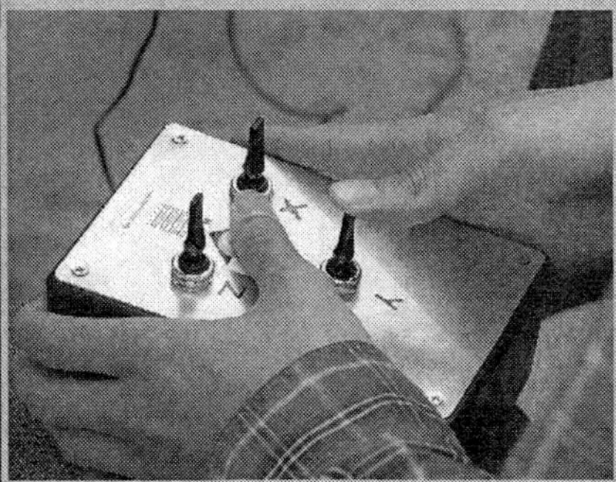

Issues:

Troubleshooting problems

www.marinetech.org

MATE underwater robotics

Troubleshooting AngelFish

Loose connections on the back of switch.

Can't see electricity flowing through wires.

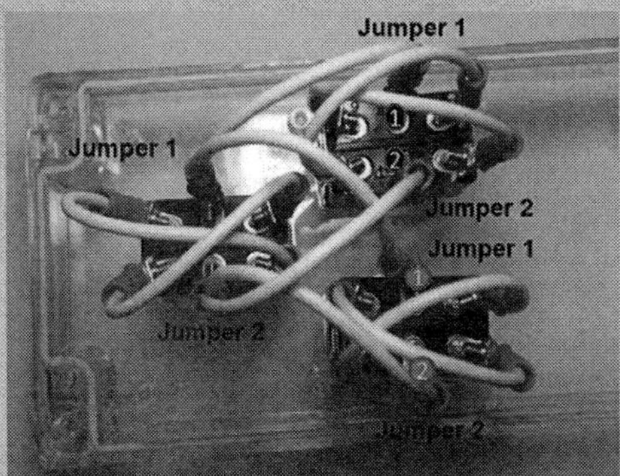

So we started adding features to alleviate the main issues and help with troubleshooting.

www.marinetech.org

MATE underwater robotics

Features added!

Printed circuit board.

- No longer soldering wires on the back of the switch. Teach the wiring as a lesson, but for the kit, it is already done.

Can't see electricity flowing through wires for troubleshooting.

- Added LEDs for power and switches.

This became a new kit!

www.marinetech.org

9

MATE underwater robotics

PufferFish ROV kit

Simple switch box controller

- Three DPDT (double pole, double throw) switches controlling three motors.

Components mounted on printed circuit board

LEDs for power in

Bi-directional LEDs for switches

Volt and amp meter included

www.marinetech.org

10

MATE underwater robotics

PufferFish Lessons

Move up from basic switches to basic electronics

- Basic electronics (LEDs, resistors, parallel versus series circuits)
- Soldering components onto a circuit board
- Voltage and Amperage

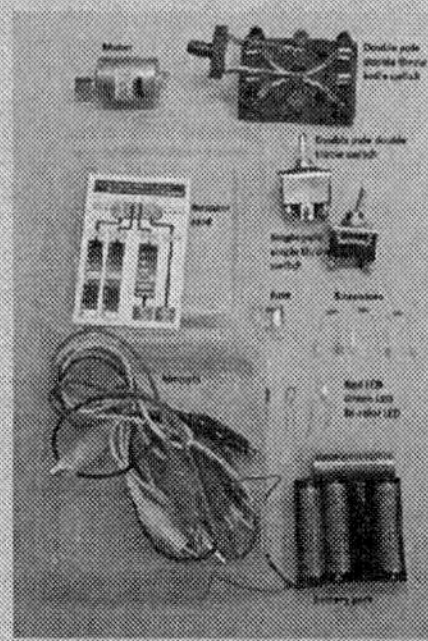

- How a DPDT switch works
- How electricity flows through a circuit
- Soldering wires together

www.marinetech.org

11

MATE underwater robotics

PufferFish ROV kit

Moved completely away from AngelFish to PufferFish for a few years.

But found that PufferFish, with a printed circuit board, electronic components, was really a beginning-intermediate ROV.

Brought back AngelFish to those who wanted or needed a true beginning kit.

www.marinetech.org

12

MATE underwater robotics

Moving Beyond Switches

Educators had learned the basics, and now wanted to move beyond beginner ROVs.

SunFish (discontinued) was too big a leap up.

Intermediate:
TriggerFish
ROV!

MATE underwater robotics

TriggerFish ROV kit

TriggerFish adds:

- Joysticks controlling multiple (2) motors.
- Analog Speed Control with no programming – Sabertooth Motor Speed Controllers.

www.marinetech.org

MATE underwater robotics

TriggerFish Lessons

Analog controls are good intermediate step:

- Motor speed controllers
- Pulse width modulation
- Joysticks/variable resistors/H-bridges

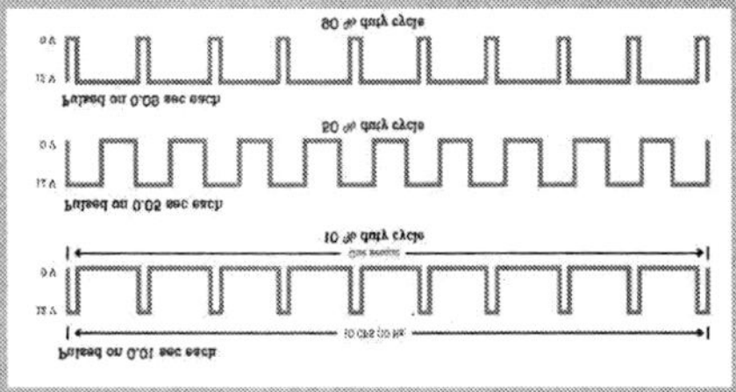

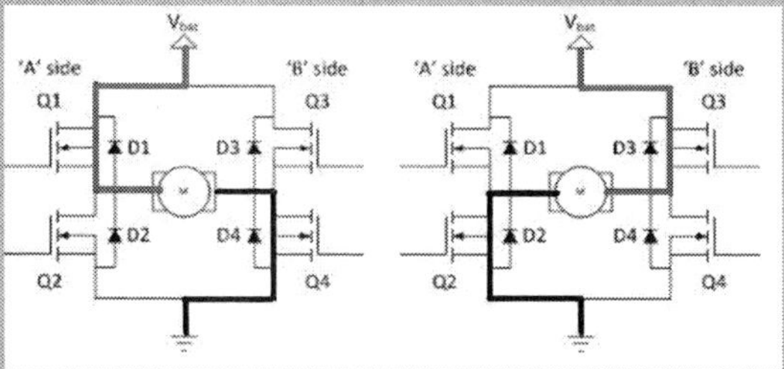

Build on previous lessons:

Basic electronics, soldering components onto a circuit board, voltage and amperage.
How electricity flows through a circuit, soldering wires.

MATE underwater robotics

TriggerFish ROV kit

Good range of motion controlled by joystick.

Easy to operate.

Knowledge of basic electronics is enough to build the TriggerFish kit.

TriggerFish: Analog controls. No programming.

MATE underwater robotics

Barracuda ROV kit

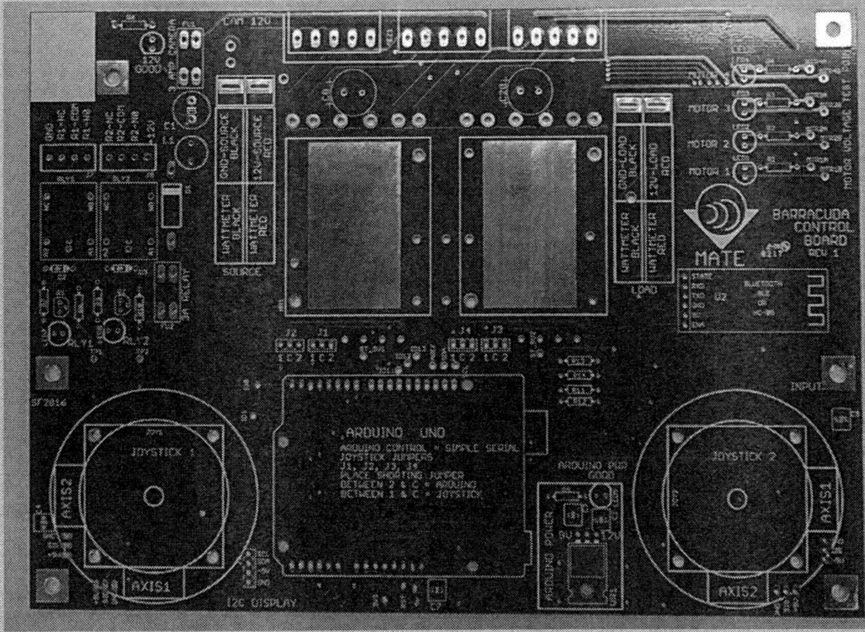

Releasing Fall 2018

MATE underwater robotics

Barracuda ROV kit

Sabertooth Motor Speed Controllers & other features of TriggerFish

- Adds Arduino
  - Arduino is used to process sensor data
  - Arduino can be used to drive if desired
  - Program “features”
- Adds Bluetooth
- Adds Relay channels
  - Actuator controls
- Output to I2C display

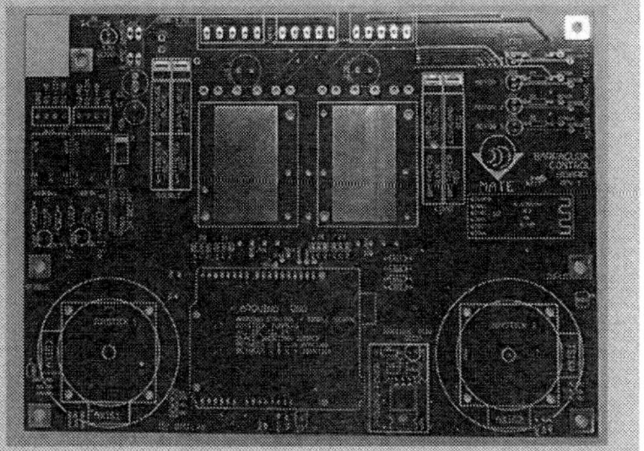

MATE underwater robotics

Barracuda Lessons

- Connecting components to an Arduino
- Programming / programs
- Sensors

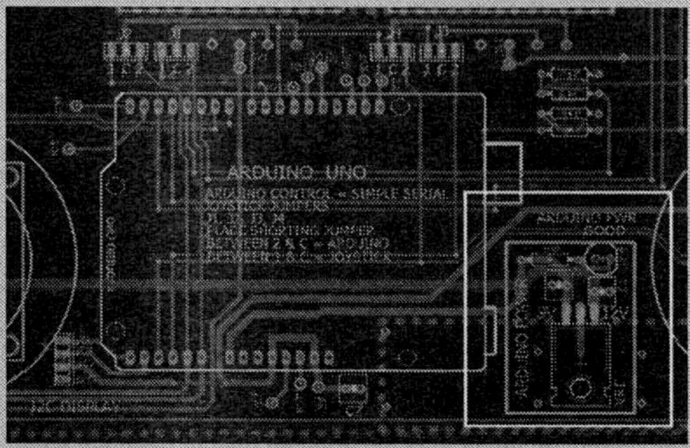

Build on previous lessons:
Motor speed control, pulse width modulation, joysticks/variable resistors, H-bridges
Basic electronics, soldering components onto a circuit board, voltage and amperage
How electricity flows through a circuit, soldering wires

www.marinetech.org

MATE underwater robotics

Moving beyond ...

Diversity of platforms available

Raspberry Pi, Beaglebone, Arduino, more powerful motors (Seabotix, Blue Robotics).

Let teams advance on their own – too many branches, too much technology, to create specific lessons.

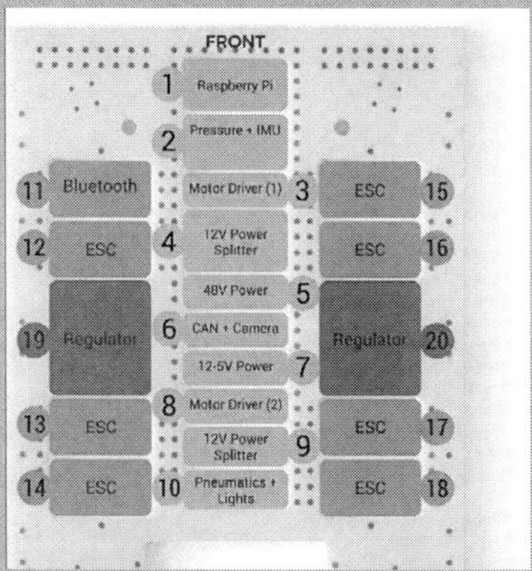

HKUST control diagram

www.marinetech.org

MATE underwater robotics

Kits Conclusion

MATE series of ROV kits allows:

- Educators and students to enter at a level comfortable to them
- Educators progress on their own terms
  - Mentor is associated with a single set of students that advances through time
  - Mentor has new students every year and stays with same kit
  - Both
- Lessons build on previous lessons without too large a leap up.

www.marinetech.org

21

MATE underwater robotics

Competition

MATE ROV Competition also has a progression of levels.

Groups can enter at a level most comfortable to them and move up as needed.

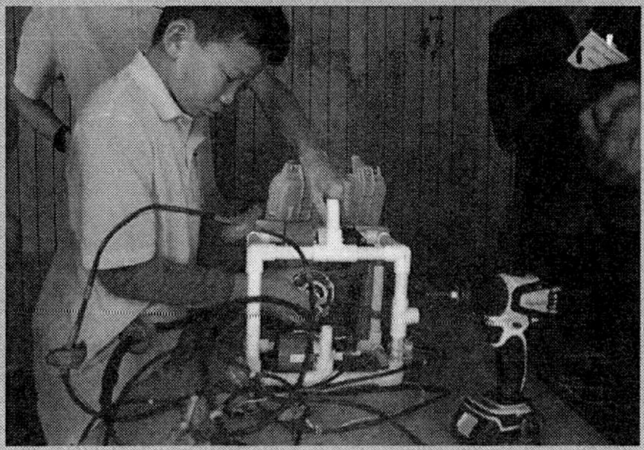

www.marinetech.org

22

MATE underwater robotics

SCOUT class

No camera needed, students can look into pool and operate their vehicle.

Simple mission tasks of picking up objects, placing objects, turning handles.

www.marinetech.org

23

MATE underwater robotics

NAVIGATOR Class

Newest competition class. MATE recognized that there was a very big step up between SCOUT (simple switch box controllers) and RANGER with computer controlled designs and sensor missions.

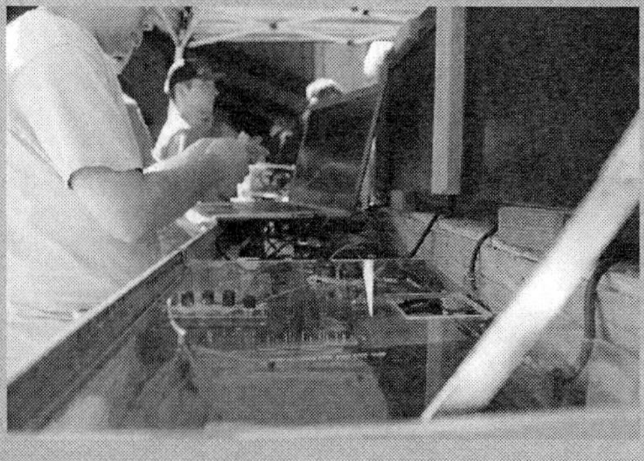

www.marinetech.org

24

MATE underwater robotics

RANGER Class

Some computer controlled ROVs. Some are TriggerFish analog joystick controls. Incorporate simple sensor tasks (temp, depth, length).

www.marinetech.org

25

MATE underwater robotics

EXPLORER Class

48 volts down to ROV. Onboard conversion. Computer control required.

Challenge with complicated tasks:

Image recognition and processing

Sensors

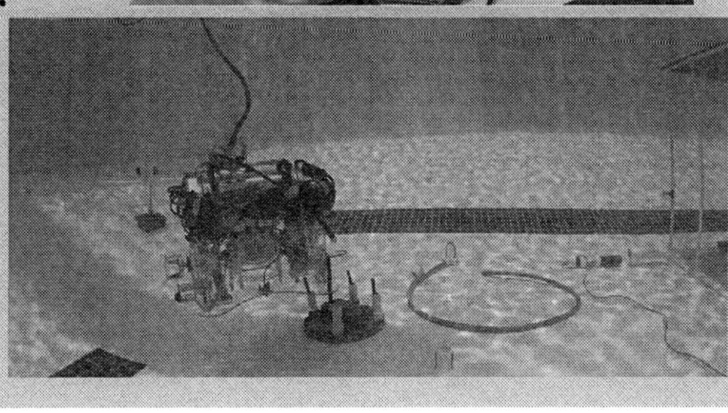

MATE underwater robotics

Competition Conclusion

MATE competition levels:

- Educators and students to enter at a level comfortable to them
- Educators progress on their own terms
  - Mentor is associated with a single set of students that advances through time
  - Mentor has new students every year and stays with same class
  - Both

Team members have moved to university and started EXPLORER class teams.

www.marinetech.org

27

MATE underwater robotics

Questions?

Matthew Gardner — Competition Technical Manager
MATE Center — mgardner@marinetech.org

www.marinetech.org

28

UNDERWATER INTERVENTION 2018

DIVING & R.O.V.

PICTURES

- PRESENTED
- BY
- BRYCE MERRILL CURATOR
- DIVING & R.O. V.
- HISTORICAL MUSEUM MORGAN CITY
- LIKE US ON FACEBOOK

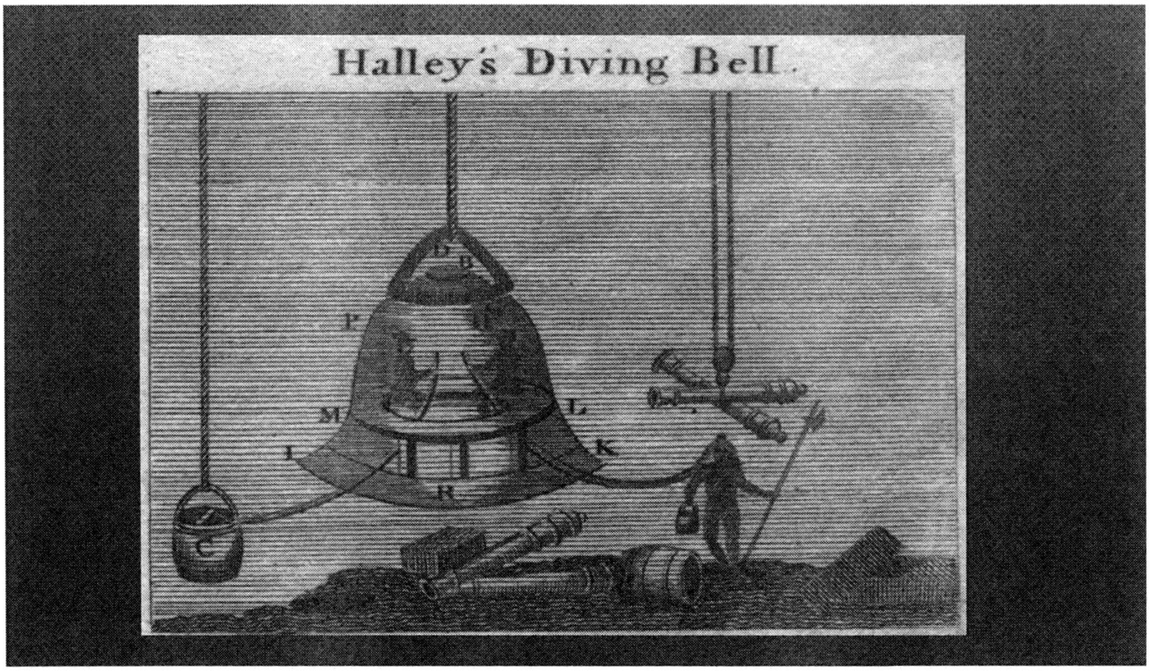
Halley's Diving Bell.

DIVERS PREPARING FOR WORK.

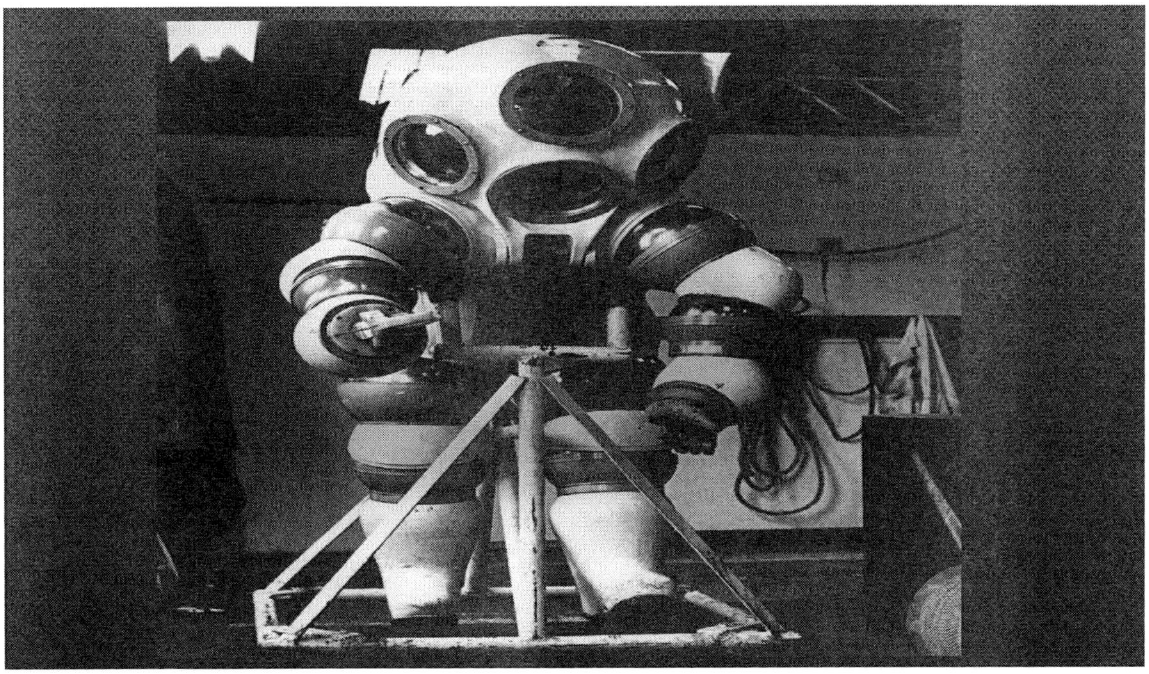

3

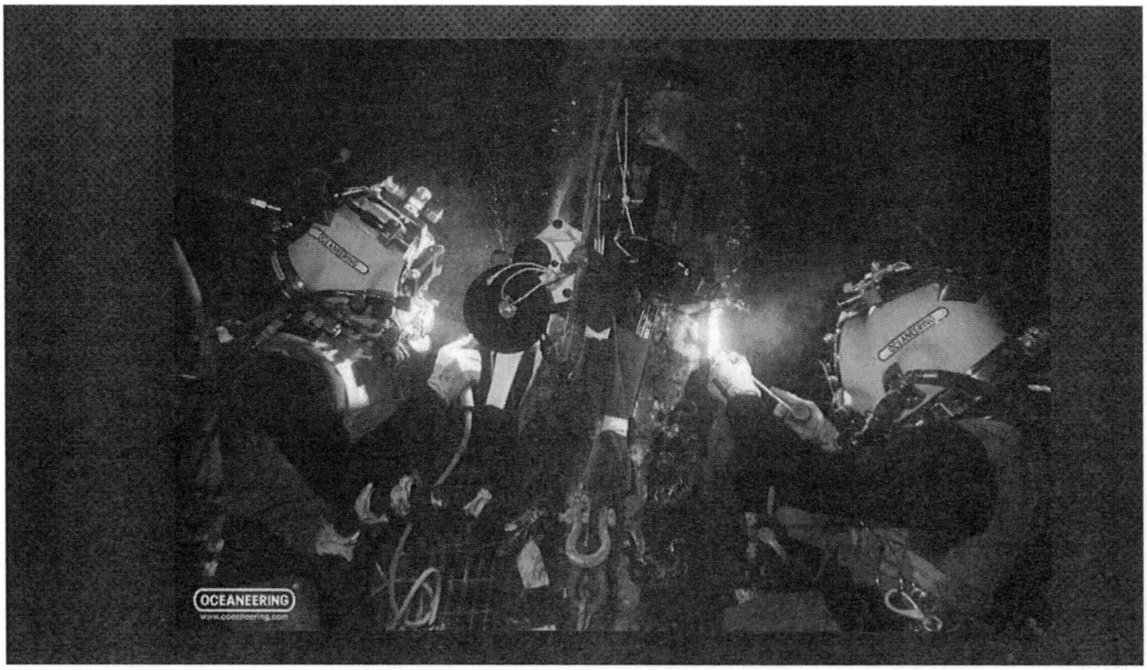
OCEANEERING
www.oceaneering.com

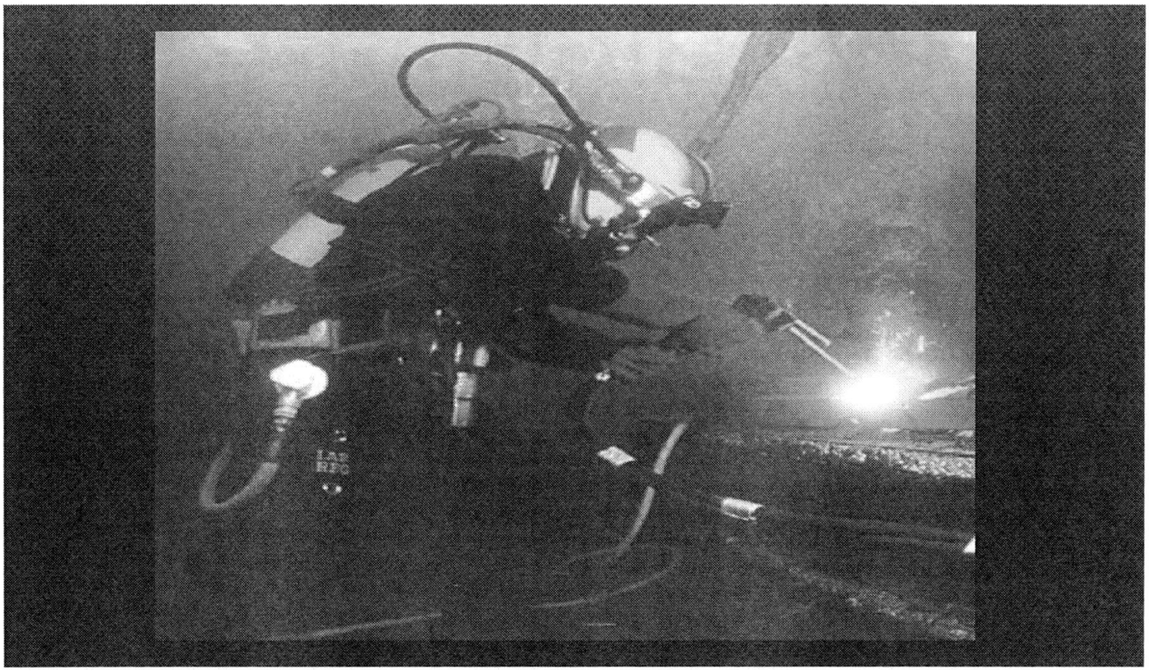

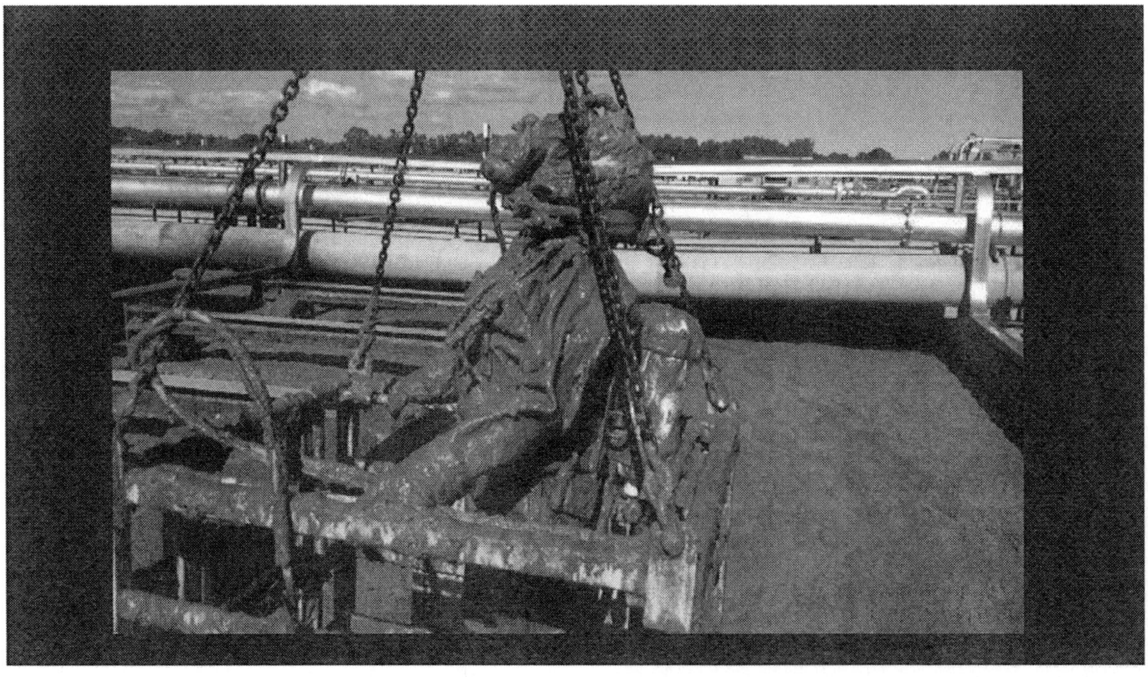

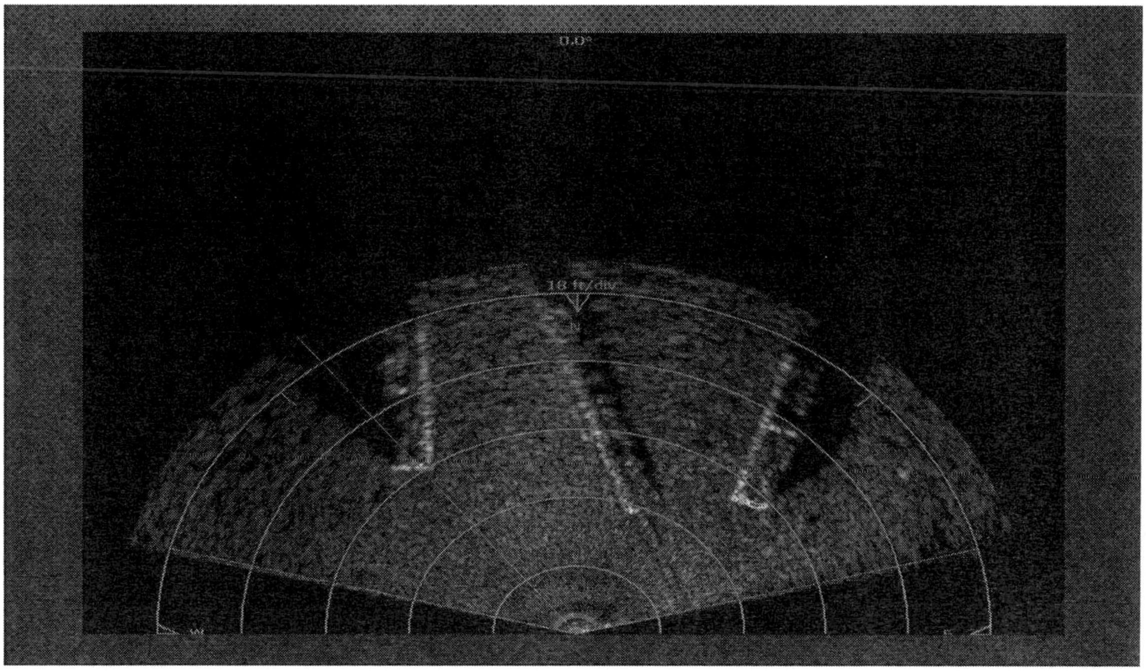
0.0°

SCARAB IV

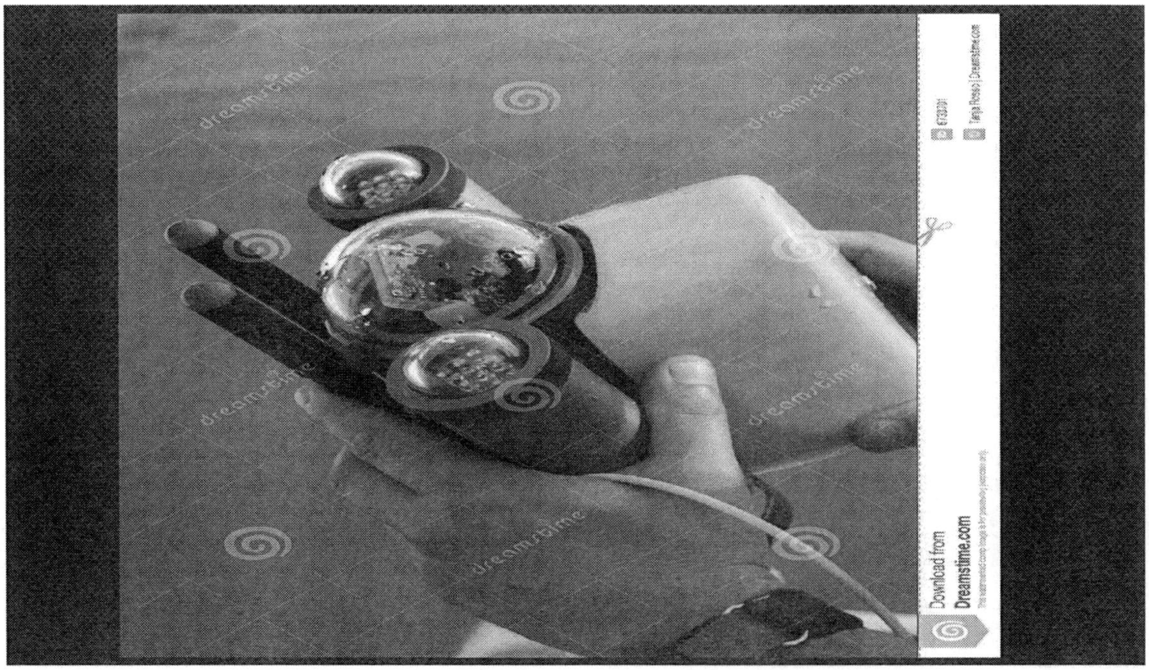
Download from
Dreamstime.com

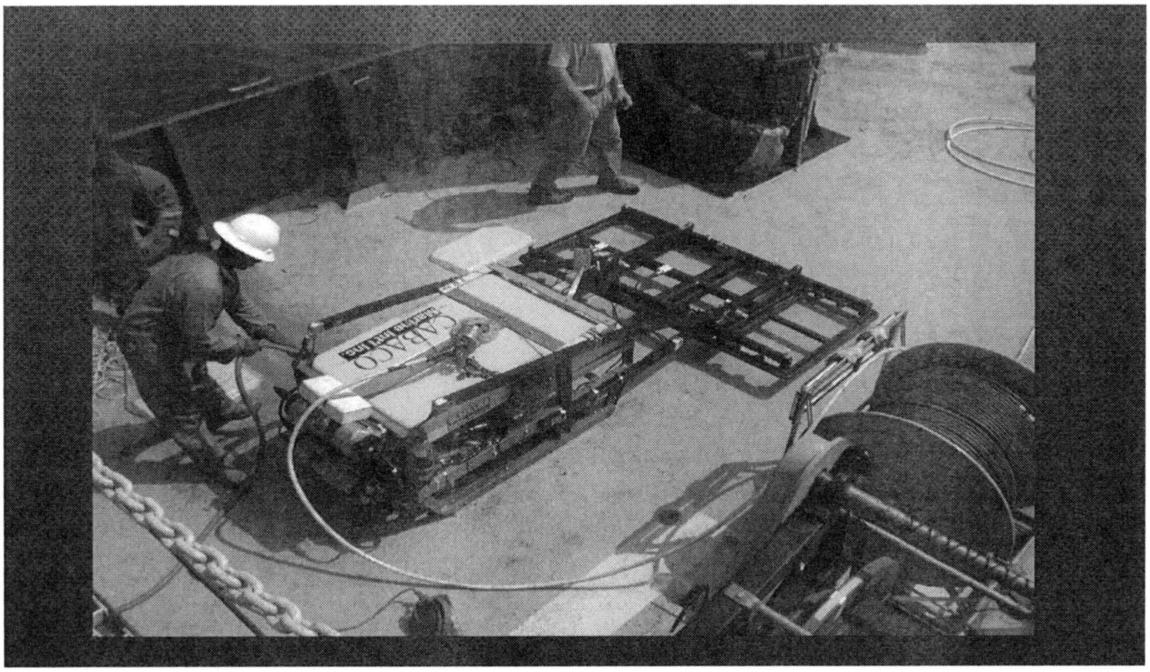
CABACO

OCEANEERING
NEXXUS

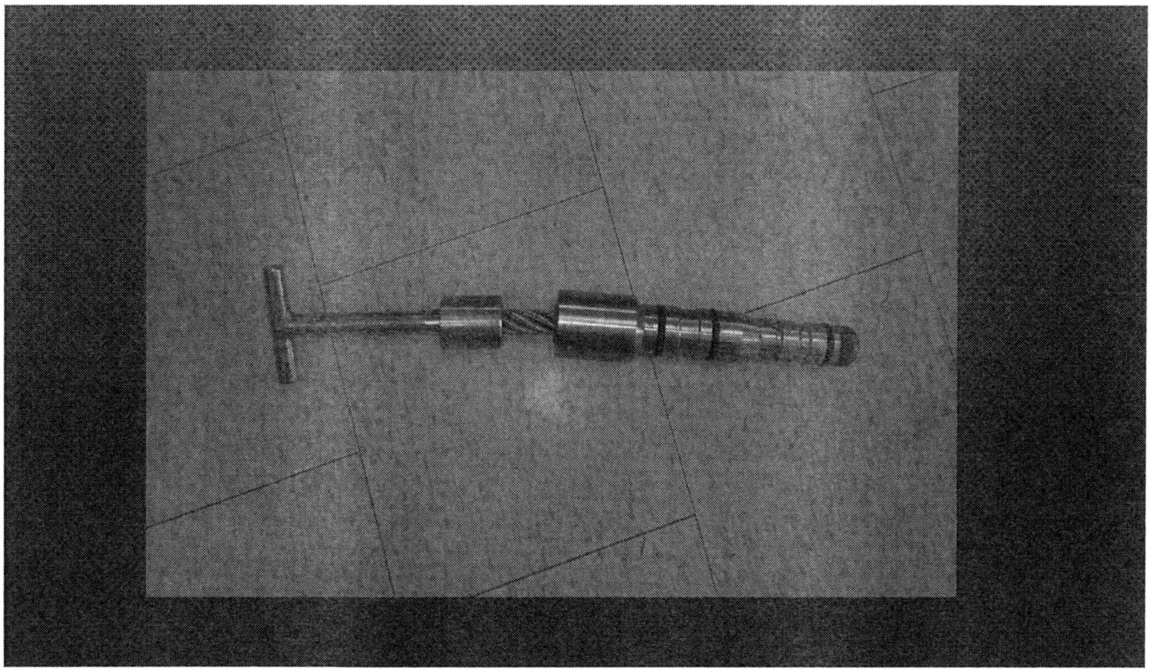

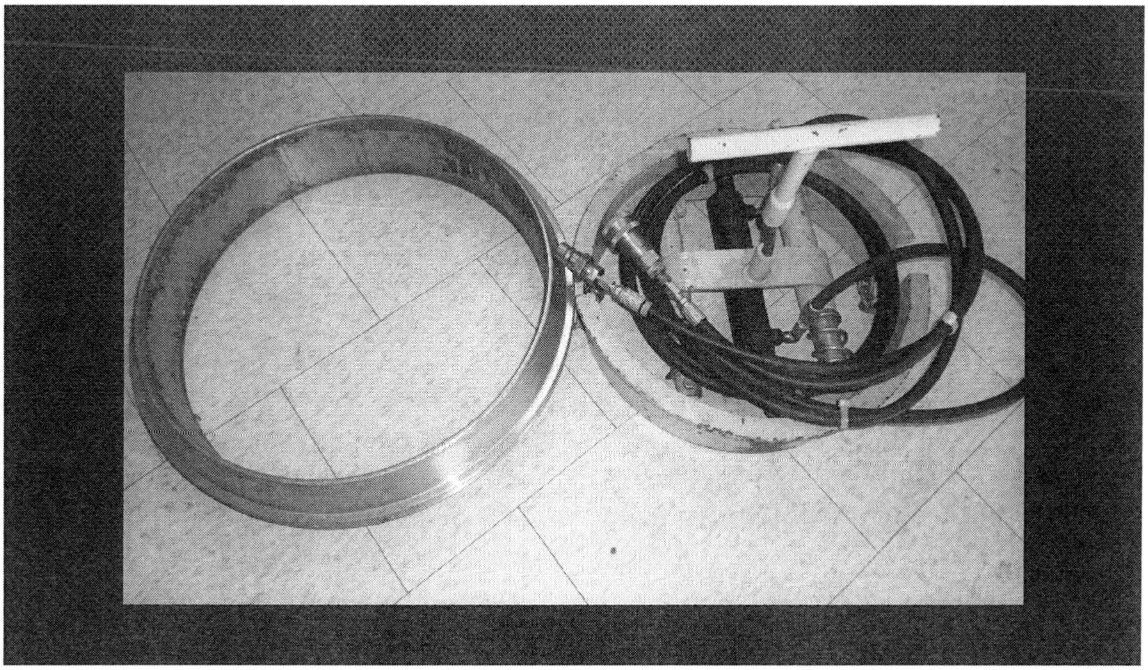

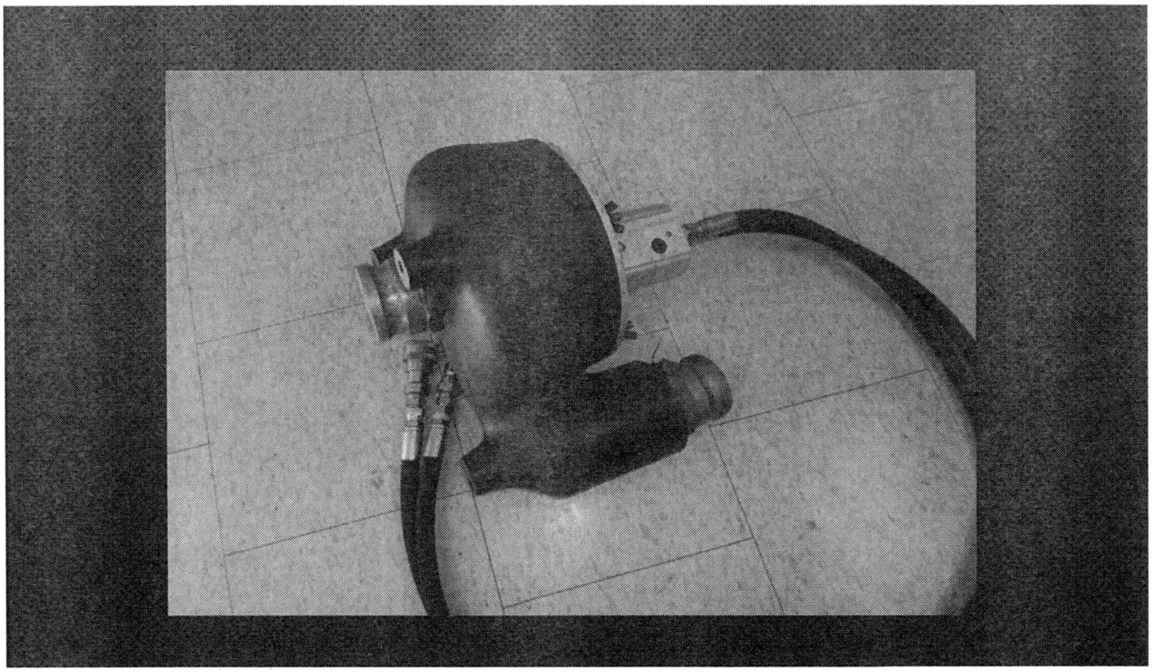

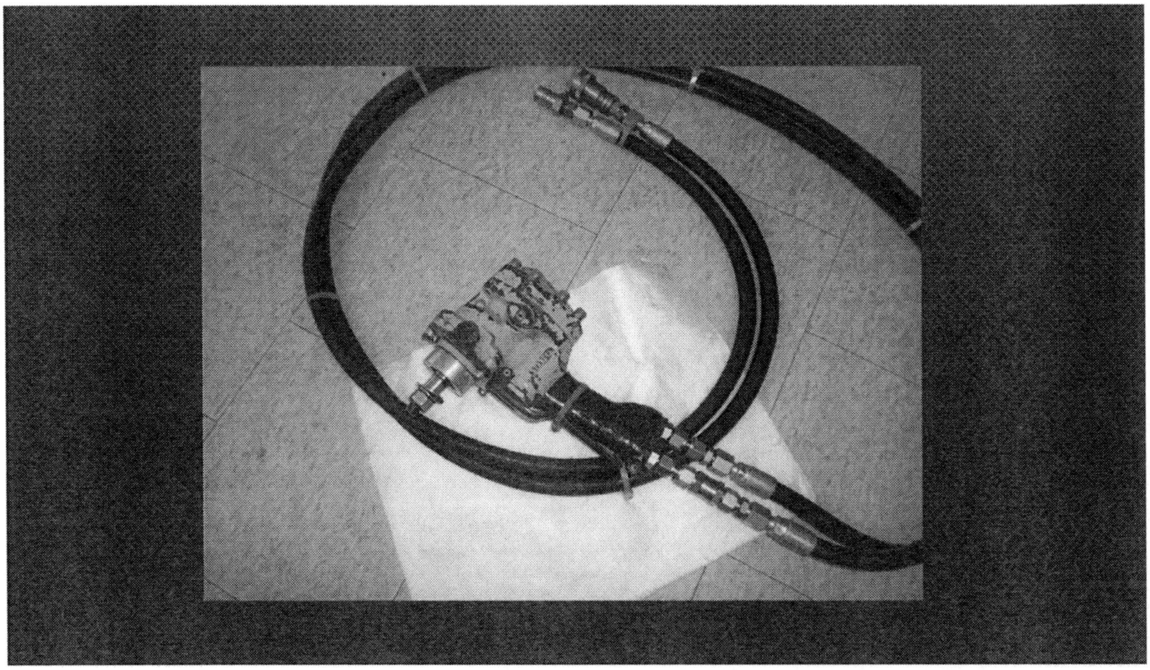

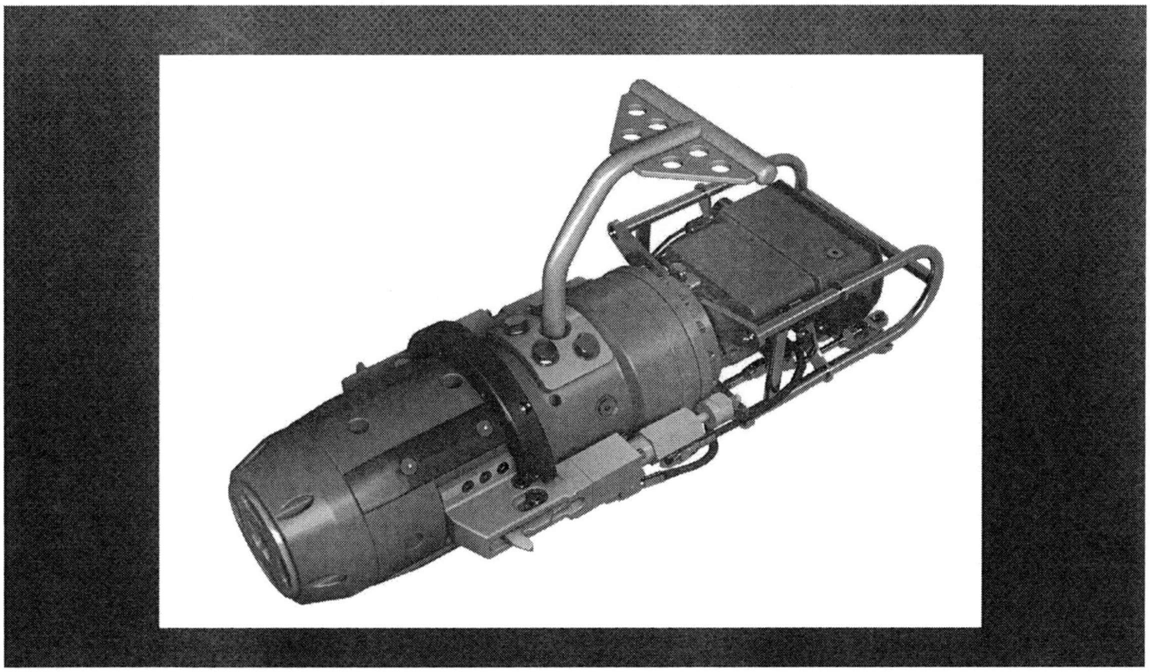

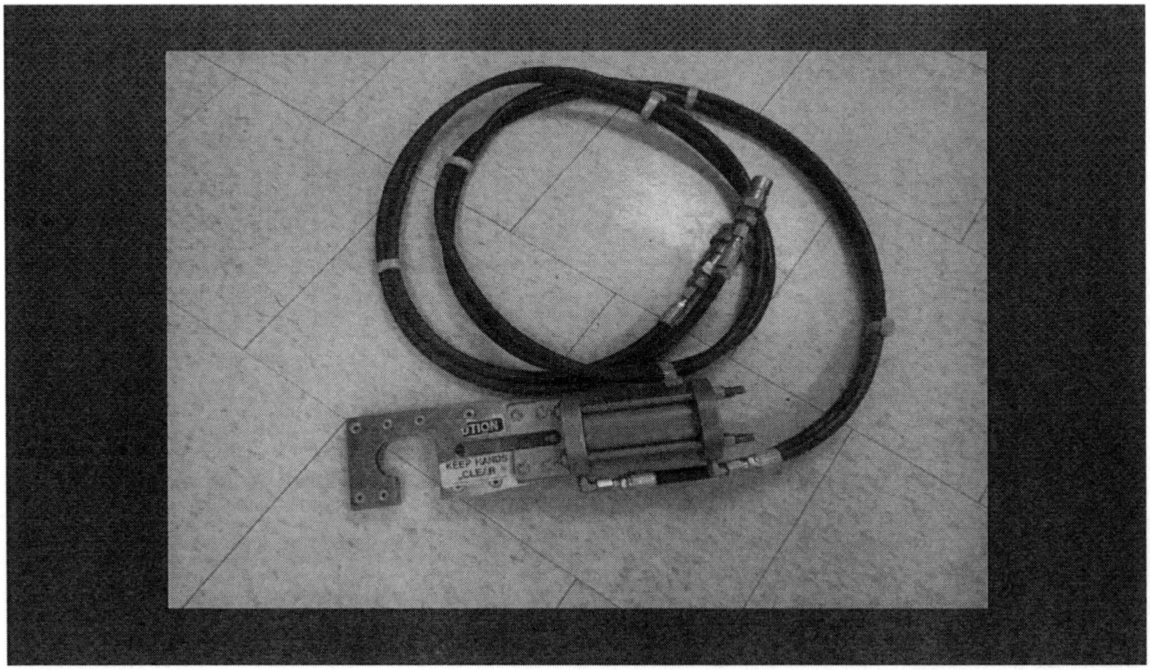
KEEP HANDS

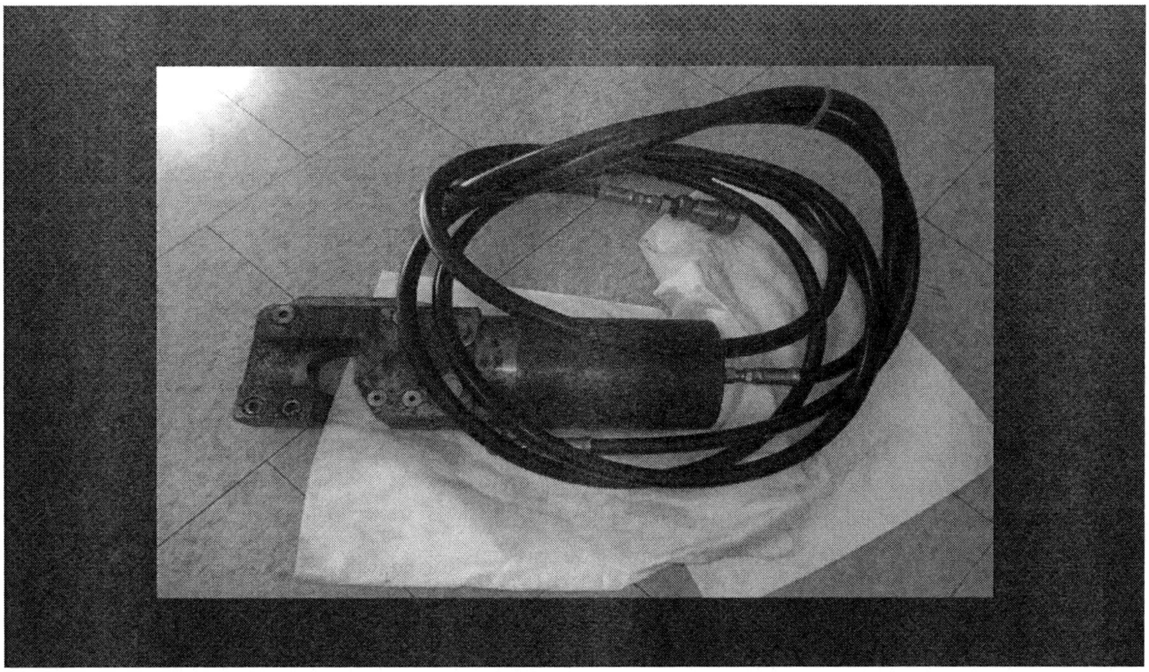

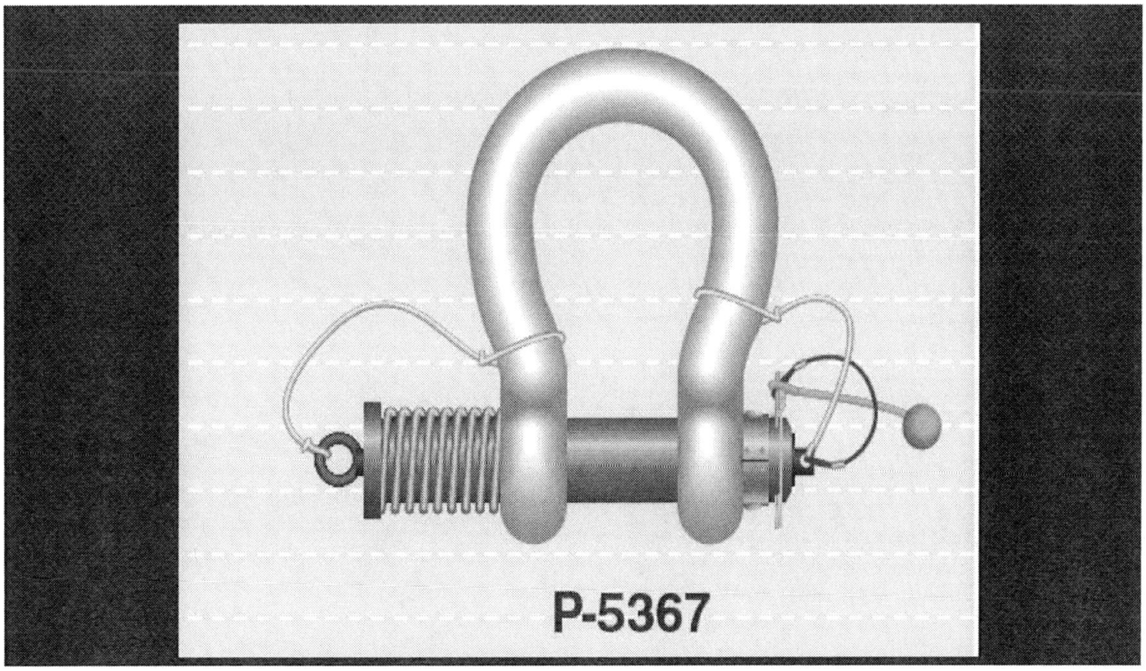
P-5367

- Thanks for you time
- I hope you Enjoyed the presentation and that it was informative.
- Questions ?

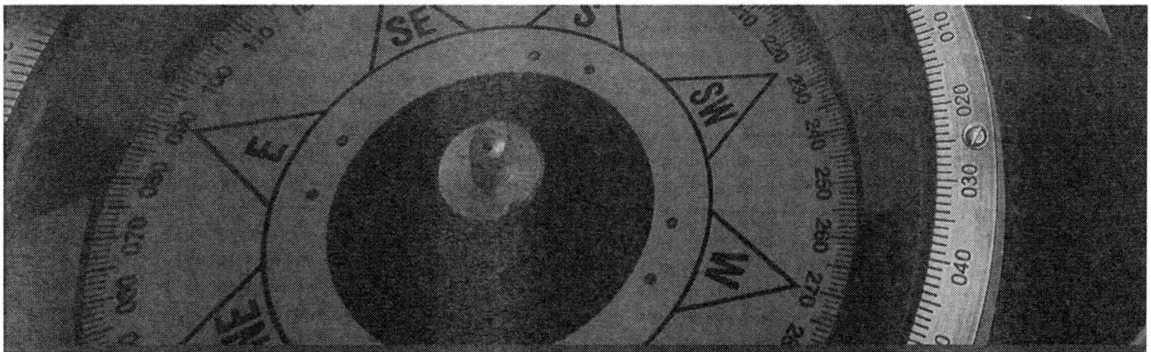

High Level Overview of Naval Mine Counters Measures Support Provided to the Royal Canadian Navy and DRDC-Atlantic.

Jim Covill[1] , Joe Podrebarac[2], Tim Dunbar[1]
[1]LR Applied Technology Group, [2]Canadian National Defence Headquarters

Lloyd's Register

Working together
for a safer world

Underwater Intervention 2018 - Presentation

Agenda

- Brief introduction to our LR ATG business unit
- Naval Mine Counter Measures
- The use of Autonomous technologies to augment more established naval mine hunting methods

Legal - Under terms of contract, LR ATG is permitted to make recommendations to the Canadian Department of National Defence for their consideration and possible further action but does not represent Canada/the Crown. This presentation does not necessarily reflect the views of the Canadian Department of National Defence

UI 2018

Introduction to LR Applied Technology Group

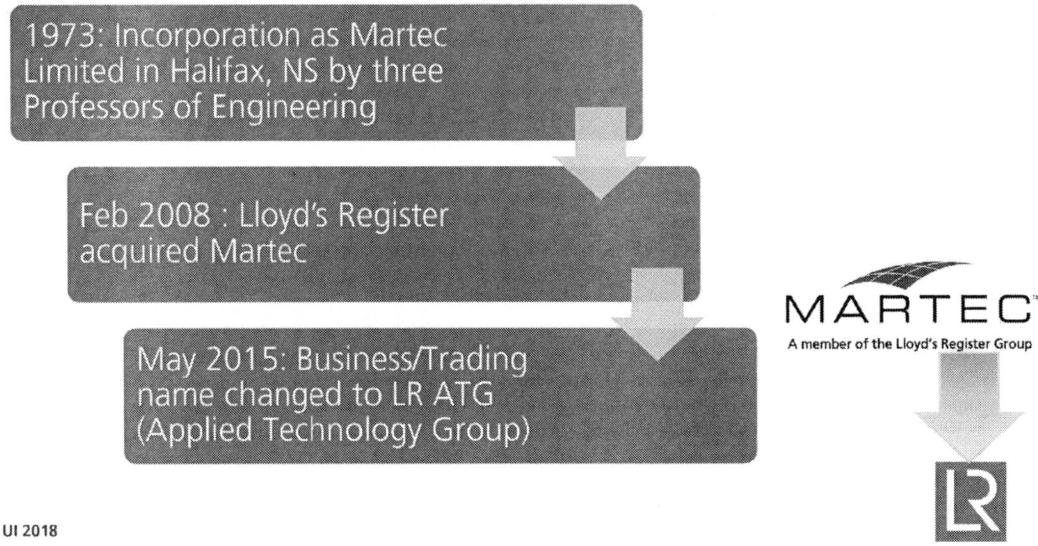

UI 2018

LR ATG – Marine Focus

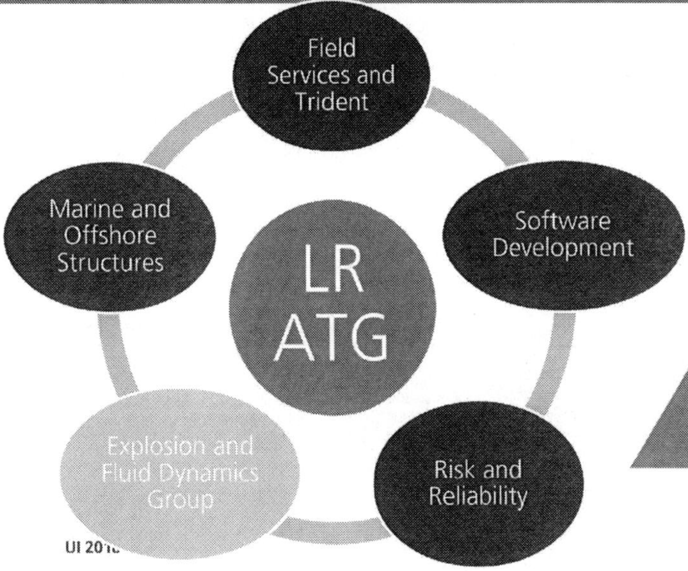

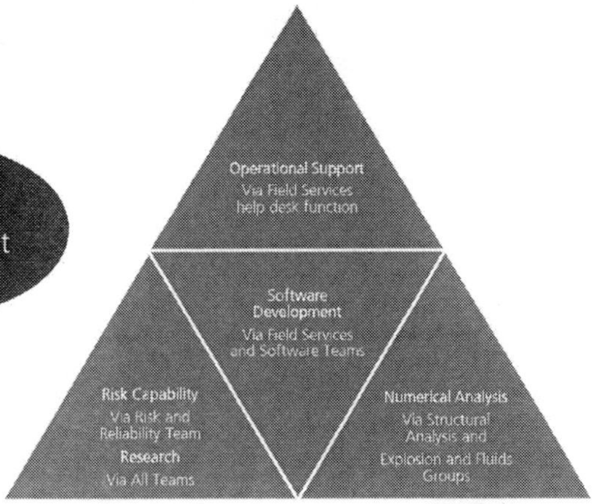

UI 2018

Introduction to Naval Mine Counter Measures

NMCM Introduction

- The Royal Canadian Navy (RCN), like many other allied Navies, is trying to improve its naval mine countermeasures capabilities (NMCM) while simultaneously trying to reduce and/or contain the costs associated with this role.
- In addition to its NMCM objective, the RCN's Fleet Diving units are often deployed to assist in *hull inspections and reconnaissance surveys*, including *underwater archaeological sites,* which permit the units to hone their skills and roles in real time situations.
- The intent of this talk is to introduce Canada's current NMCM focused capabilities and thought processes to the UI 2018 community with the goal to solicit commercial feedback.

HMCS Saskatoon

Introduction to Mine Warfare

Mine Warfare in traditional form is two disciplines:

1. Mining
2. Naval Mine Countermeasures (NMCM).

NMCM includes all measures for countering sea mines by reducing the risk or preventing damage to ships and personnel and is further subdivided into two main areas:

Offensive NMCM - proactive measure to eliminate an enemy's mine-laying capability prior to them undertaking such hostile action.

Defensive NMCM is broadly focussed on conduct of operations to deal with mines once laid, or are suspected of being laid, by identifying mine laying activity and reducing the threat and is broadly divided into Passive and Active NMCM.

UI 2018

Introduction to Mine Warfare ...

Passive used to localise the threat include establishing routes, establishing diversions, closing ports, using accompanied shipping, and providing navigational warnings.

Active involve the operation of sensors and neutralization systems, whether air, surface, or sub-surface mounted, to locate and dispose of sea mines. This includes the process of detection, classification, identification, and marking of contacts for eventual disposal.

Route Survey is traditionally considered a passive MCM technique used as battle-space shaping technique in order to prepare the area for the event of future live operations. But when RS is undertaken because of evidence of a possible mining operation and a mine is then found, it effectively becomes active NMCM in the form of mine-hunting.

UI 2018

Introduction to Naval Mine Counter Measures

Naval Mine Counter Measures operations are often conducted in four successive stages:

- Detection: Typically utilizing some type of sonar. Sonar is used to search the seabed/water column in order to detect potential 'foreign objects'.
- Classification: Techniques are then employed for each new object encountered (e.g. second look higher frequency sonar, ATR ...). Depending on the environmental conditions, mine(s) anticipated and operator's skills, contacts are classified as either mine or non-mine like.
- Identification: Objects classified as mine-like, warrant closer look via additional means deployed to visually identify the contact (e.g. optical/video sensors on an ROV/AUV/ATR or clearance divers).
- Neutralisation.
- Introduction movie (Why we do what we do!) CLIP

UI 2018

LR ATG involvements in Naval Mine Counter Measures

Our involvement in NMCM includes:

- Operational Support including Risk Assessments
- Research & Development
- LR can provide an independent Third Party perspective to DND, RCN and Naval Engineering Test Establishment (NETE)

RCN's Use of Inspection Class ROVs

RCN has gained valuable experience/usage from these ROVs over the past six years.

Sample surveys undertaken have included:

- Submarine torpedo hatch inspections
- Underwater/archaeological ship surveys in Canada's Arctic waters
- Capsized fishing vessel survey (body search mission)
- Mine-like bottom object inspection (typically undertaken during naval exercises)
- Canadian WW II sunken aircraft recovery including skeletal remains
- Jetty inspections

'Miss Ally' Recovery Mission – February 2013

Operational Support

NMCM Software Support Capability

Operational Support

Domain Support:

- Mapping/Navigation/GPS
- SSS Data Acquisition
- SSS Data Post Processing
- Bathymetry Post Processing
- ROV software
- Database (for Mines)
- CFD Analysis
- FE Analysis
- AUV software development
- Various other software as needed.
  - Agisoft Photoscan
  - ...

UI 2018

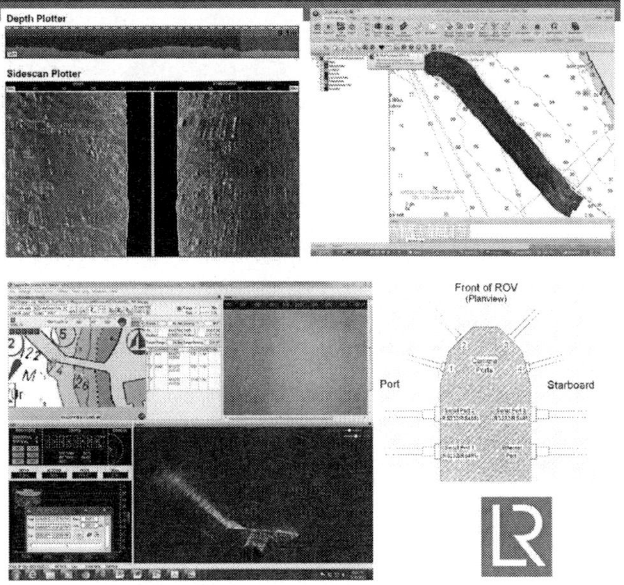

Support of Remote Operated Vehicles and Side Scan Sonar

Operational Support

Software Support
- Investigate issues; assist; trouble-shoot
- Interface with client and/or manufacturer
- Standardize software installations

Quick Reference Guides
- Develop quick setup instructions
- Develop guides to quickly perform a function
- Setup FTP secure server to facilitate distribution

Demonstrations
- Demonstrate software (with presentations & exercises)
- Investigate/learn tools and then demonstrate how best to use them

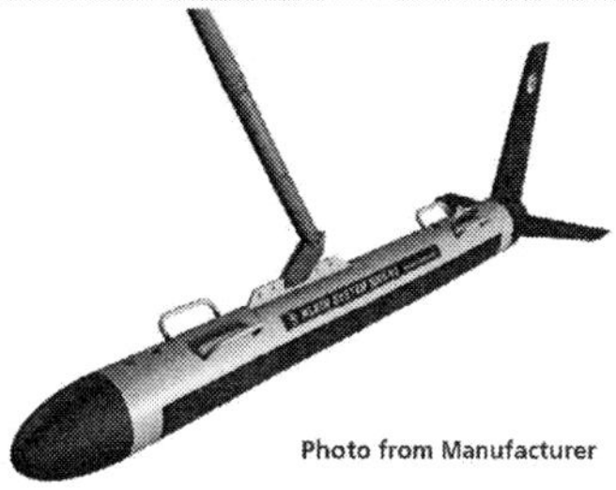

Photo from Manufacturer

UI 2018

Charts Support (including GPS related)

Operational Support

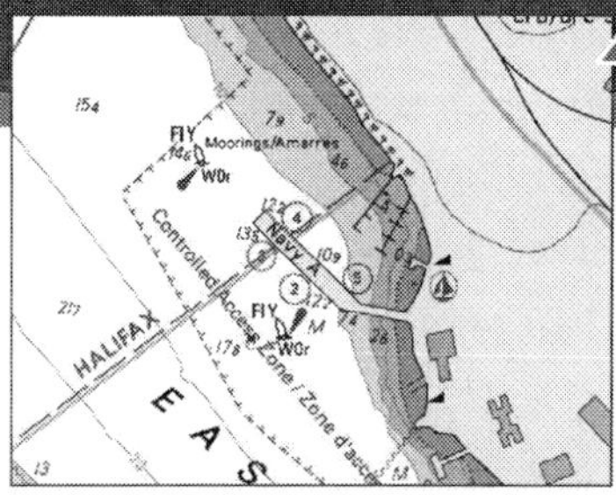

Support Software
- Global Mapper, WinFrog, Projection Conversions, GPS

Create Quick Reference Documents:
- "Global Mapper Basics"
- "WinFrog Basics"
- "UTM Basics" (+geographic co-ord. system/projections)

Obtain Maps
- Canadian Hydrographic Service
- Obtain charts for the Arctic (with bathymetry not typically available)

Create / Geo-rectify Images

Side Scan Support Capability

Operational Support

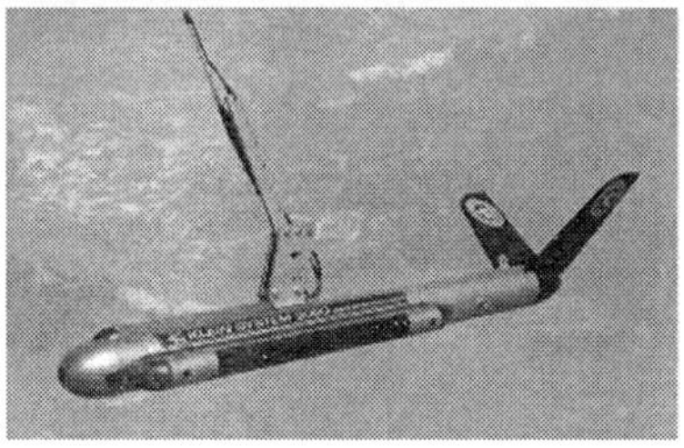

Support Software

- Scanline, Sonar Pro, SonarWiz, Fledermaus

Create Quick Reference Documents:

- "Quick Guide for Klein Side Scan Sonar Data Acquisition Using SonarPro"
- "Quick Guide for Klein Side Scan Sonar Data Acquisition Using SonarWiz"
- "SonarWiz Basics" (Post-processing)
- "Introduction to Fledermaus Software for Post Processing Bathymetry"

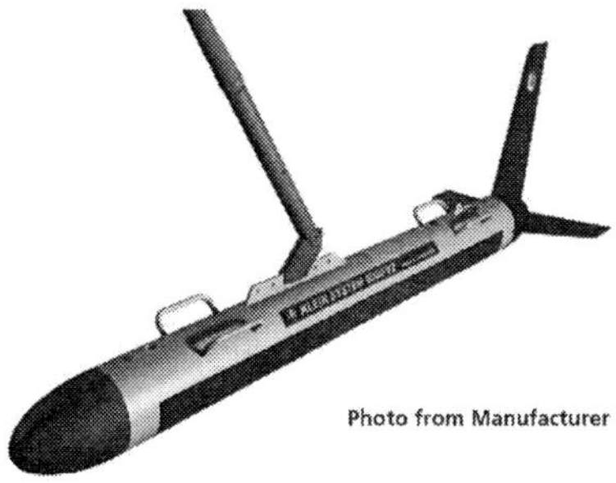

Photo from Manufacturer

UI 2018

ROV Support Capability

Operational Support

Support Software

- SeaNet Pro
- Data Manager
- SPS Programmer
- Port Manager
- Firmware / Firmware Manager

Create Quick Reference Documents:

- Add/Remove Charts
- Standardize Software Versions/Setup
- Com-setup
- Quick Component Description / Wiring / Troubleshooting

UI 2018

Software Development: Mine Database

Develop a Mine Database (MIDAS)

- CF Maritime Warfare Centre house MIDAS
- ATG was asked to remedy issues with current database
  - Create/populate a new database
  - Create a user-friendly GUI (Interface)
  - Add database filtering mechanism
  - Classification: NATO Secret

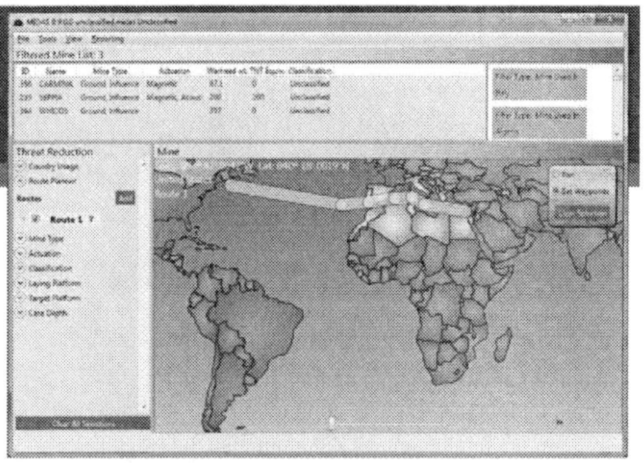

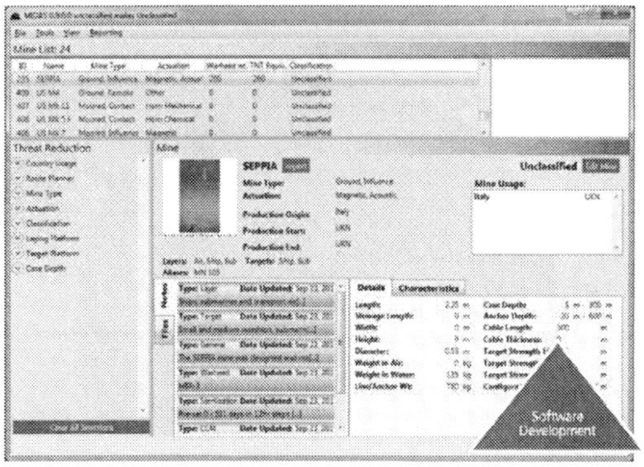

UI 2018

Software/Hardware Investigation: DSIMS - VEMS

Research

Investigate / assess the capabilities of the VEMS and DSIMS systems

- VEMS: Versatile Exercise Mine System
- DSIMS: Diver Signature Integrated Measurement System

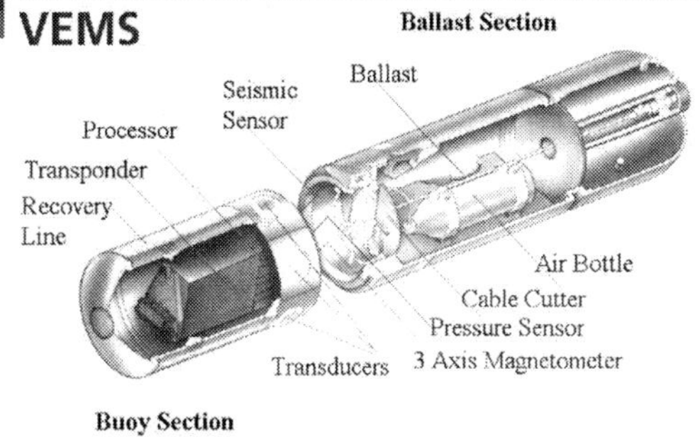

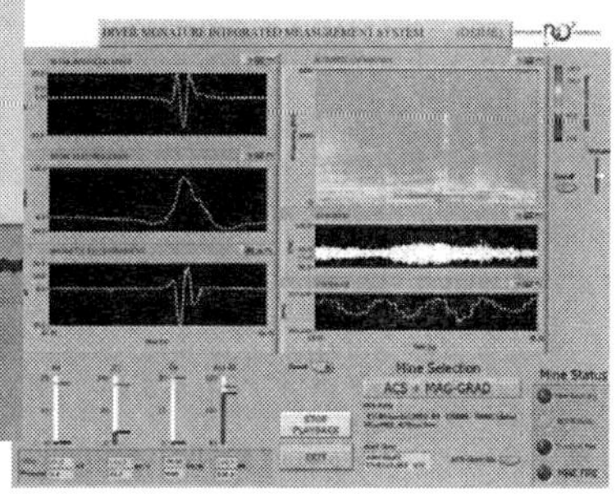

UI 2018

Numerical Support: Finite Element & Computational Fluid Dynamic Analysis

FEA/CFD analysis of a human skull

- Ultimate Goal: Promote diver safety
- NMCM contract was a mechanism for numerical analysis
  - Developed a detailed model of a human skull
  - Performed dynamic structural analysis
  - Real explosives applied to this model in the Halifax Harbour
- DRDC (Toronto) still continues work on this model today
- Sensitive material (US data)

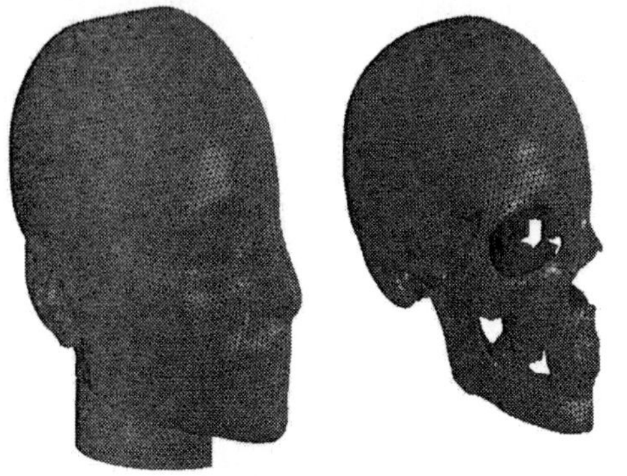

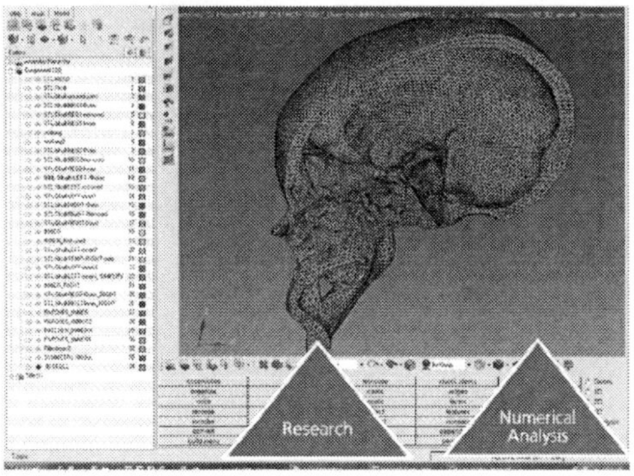

UI 2018

Numerical Support: Structural Analysis

Structural Assessment of a LARS (THD)

- Perform global analysis (essentially beams)
- Use global reactions to perform checks on the members.
  - Tension, Compression, Bending, Shear, Combined
- Check Welds and Pad eyes
- Check Reactions at 4 ISO Container connections
- Sensitive (Controlled Goods)

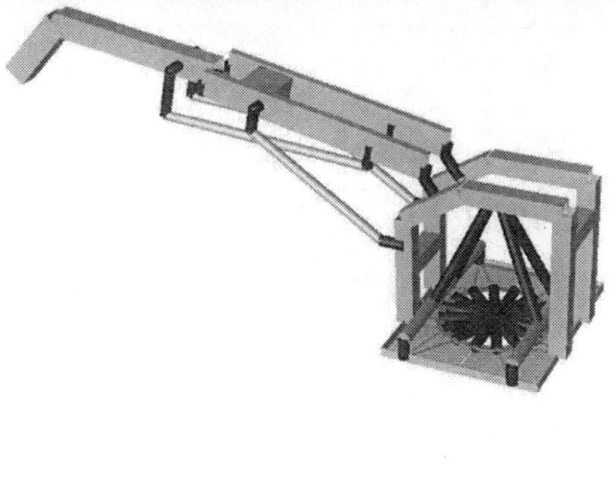

UI 2018

Extensive Software Development for AUVs (Topside and Internal Software)

Software Development

Develop AUV Mission Planning Software for DRDC

- Consisted of creating SeeTrack Add-ins / Plug-ins
- Define survey regions using SeeTrack
- Execute an add-in (we created) to generate a mission file
  - Reads / writes to SeeTrack database
  - Requests user input (survey region, start/stop points, vehicle – IVER)
  - Calls a DRDC created program (PATHA) to determine optimal track spacing
  - Creates/displays a SeeTrack mission file
  - Creates IVER specific mission files (*.MIS and *.SRF)

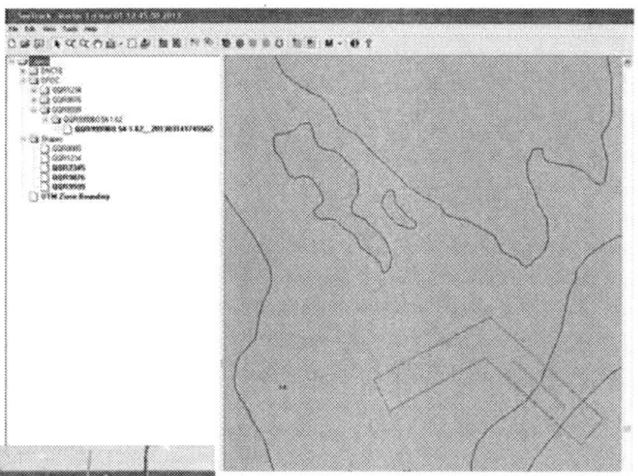

UI 2018

Current Activities: Autonomous Technologies

Research

Operational Support

Evaluate autonomous technologies

- Assess the control/communication of autonomous technologies with divers in mind (operational)

Attend Pax River and Unmanned Warrior Trials

- Objective: Bring together operators, developers, and stakeholders to jointly explore AUV technologies in a common at-sea environment.
- Multi-nation trial (AUV/USV)
- Gain a users-perspective on REMUS and IVER AUVs
- See Neptune in action with ASV and AUVs
  - Reacquire behaviour (by alt. AUV) worked!

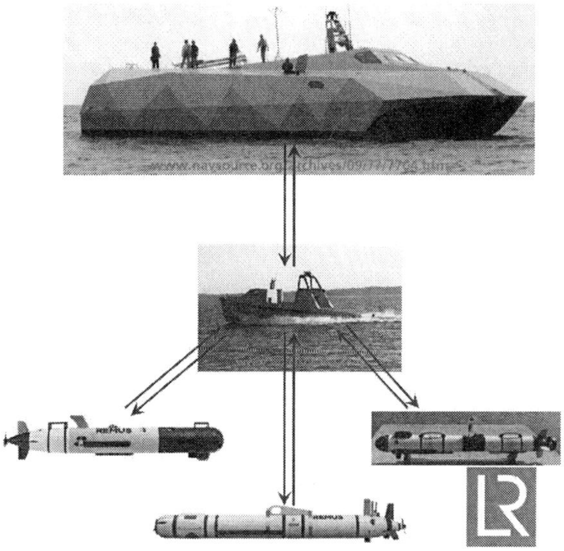

AUV Photos from Manufacturer

Extensive Software Development for AUVs (Topside and Internal Software)

Software Development

Vehicle Interfacing
- Develop software to convert mission files to vehicle specific formats (IVER, Arctic Explorer, general SeeTrack format)

Behaviour Development
- Develop specialized 3rd party behaviours for expansion on autonomous control systems
- Ex. Waypoint Reacquire, Filter Sonar File, Adding 3rd party developed ATR functionality

Vehicle System Monitoring
- Develop integrated monitoring functions for run-time tracking of autonomous vehicles.
- Ex. Monitor depth, turning, ascending, descending, files

UI 2018

LR

Integration of Technologies

Operational Support

Training Operation

- Following turnover of two experienced senior staff
- Hunting for an 'Object' in shallow water
- Ideal for small boat and Starfish SSS
  - Practical search and find tool;
  - Easily deployable tool;
  - Can be used on most boats
- Create maps
- Practice setup
- Practice data acquisition
- Practice post-processing

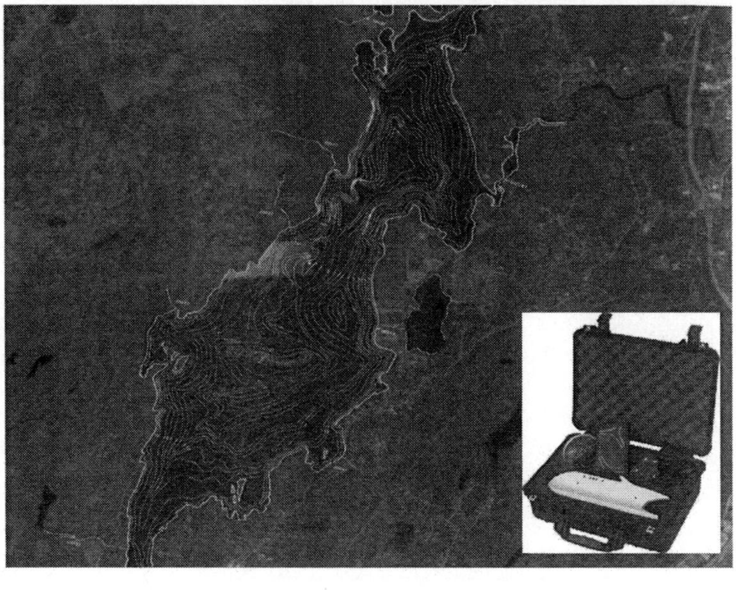

UI 2018

Current Activities: Autonomous Technologies

Operational Support

In 1943, a Harvard (AT6) aircraft crashed in a training exercise into a lake in Nova Scotia.

- ~88 year old eye witness (narrow search region)
- Pilot survived the crash, which means no human remains exist which simplifies the exercise.
- Objectives:
  - Assess USV capabilities (Bathymetry, GPS, Wi-Fi…)
  - Assess AUV capabilities (SSS, INS, ACOMMS…)
  - Conduct SSS and ROV investigations
  - Observe the top down data flow

Current Activities: Autonomous Technologies

Operational Support

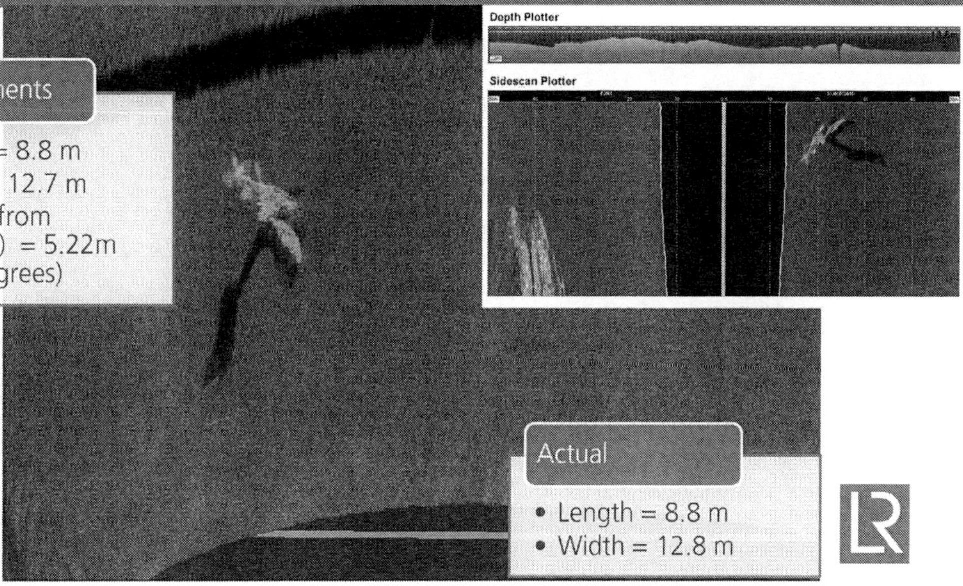

Measurements

- Length = 8.8 m
- Width = 12.7 m
- Height (from Shadow) = 5.22m (~35 degrees)

Actual

- Length = 8.8 m
- Width = 12.8 m

Current Activities: Autonomous Technologies

Operational Support

Measurements

- Length = 8.9 m
- Width = 13.1 m
- Height (from Shadow) = 5.33m (~35 degrees)

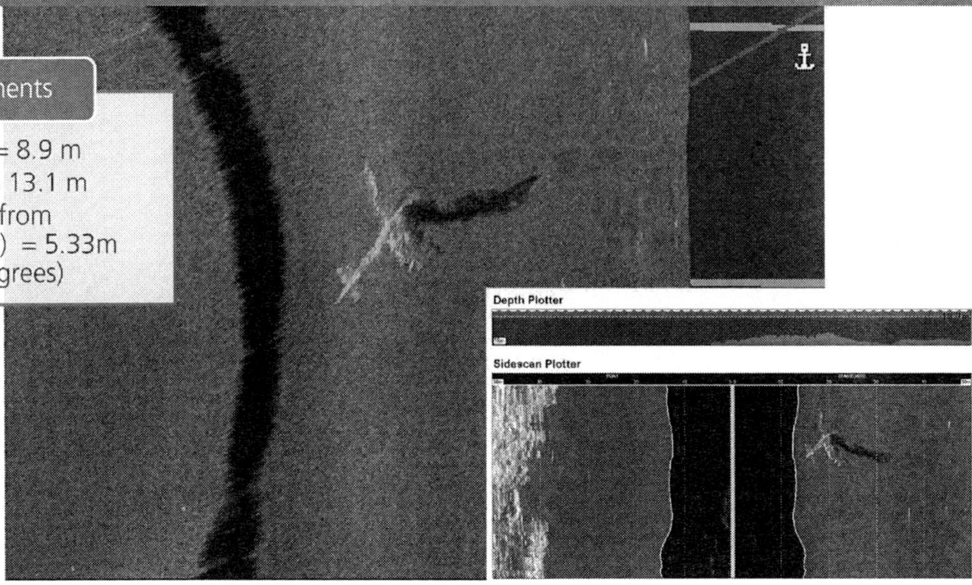

UI 2018

Current Activities: Autonomous Technologies

Operational Support

ROV

- Launched ROV at position found by the SSS
- Plane quickly appeared on screen (Multi-beam sonar image)

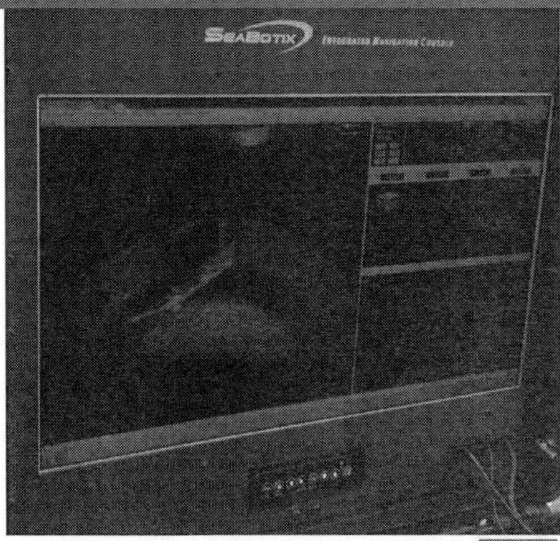

Current Activities: Autonomous Technologies

Research

Operational Support

Actual Exercise @ DRDC's Mine Garden

- Capability demonstration exposure to operational units
- Three AUVs: IVER 2 and 3 and REMUS 100
- Investigate user intentions / operating envelopes:
  - Pre-trial survey: see what the naval community wants
  - Trial: Exposure/Investigate what systems can do
  - Post-trial survey: see user likes/dislikes, ease of deployment, use in the field...
- Investigate performance
  - Compare sonar performance, & ATR with different sonars
  - Compare navigation on AUV platforms both using a Kearfott T16 INS

http://www.whoi.edu/oceanus/feature/where-are-mines-hiding-on-the-seafloor

Future Role of Autonomous Technologies in NMCM

Research

Software Development

Canada, like our allied navies, is actively funding research, development and end user delivery of autonomous technologies including UUVs and AUVs for both bottom surveys and mine neutralization.

Looking to the future, autonomous vehicles should be able to provide a far more cost effective and safer means of conducting underwater surveys in hostile environments.

Canada is currently planning on acquiring AUVs and finalizing its Remote Mine Disposal System (RMDS) to take advantage of autonomous technologies.

The RMDS will be a modular, stand-off Naval Mine Countermeasures capability, designed to provide a full spectrum of Naval Minehunting operations and contribute to underwater domain awareness.

Remote Mine Disposal System

As currently published (2017-18) Canada is looking to acquire a system envisioned as:

1. AUV – Search, Detect and Classify subsystem:
   - Minimum four (4) Man Portable Shallow Water AUVs (not more than 45 kg)
   - Minimum four (4) Light Weight Deep Water AUVs (up to 300 kg)

2. Mine Disposal Subsystem:
   - Minimum four (4) Inspection/Training variant - unarmed reusable
   - Minimum twenty (20) Combat variant – explosive ammunition capable

3. Transportable Command Centre Subsystem:
   - Minimum two (2) Standard 20' ISO support containers

Remote Mine Disposal System

In addition, Canada has funded independently of the above system, the following research related to NMCM AUV initiatives:

1. Computer augmented AUV mission planning and mission execution,
2. AUV behaviour including mission error recovery logic
3. Automatic target recognition/identification
4. Fault Detection and Recovery
5. Risk Assessment

Conclusions

After nearly six years of providing support, one of the more interesting observations is simply the dynamics of technology :

- Expectations have grown with a newer/younger PC savvy generation of RCN personnel
- Inherently software tends to become more complicated as newer versions are released.
- Basic functionality can change in subtle ways
- New 'bugs' and problems crop up from time to time

UI 2018

Conclusions ...

- Individual personnel favour different ways of doing things
- Data assimilation remains a challenge between the various products
- Frequent training/coaching due to RCN redeployments

UI 2018

Conclusions

- RCN personnel have come to rely on new equipment and technologies and are not content to remain status quo. At the same time, we are constrained by budgetary realities, a requirement to maintain operational status while introducing hardware and software upgrades, and new equipment into the RCN.

- Clip

UI 2018

Jim Covill
Field Services and Trident Group
T 902 407 9788 E jim.covill@lr.org

Joe Podrebarac
DNCS 4, Canadian Department of National Defence
E joseph.podrebarac@forces.gc.ca

Lloyd's Register

Working together
for a safer world

Growing a marine technology program for middle and high school students

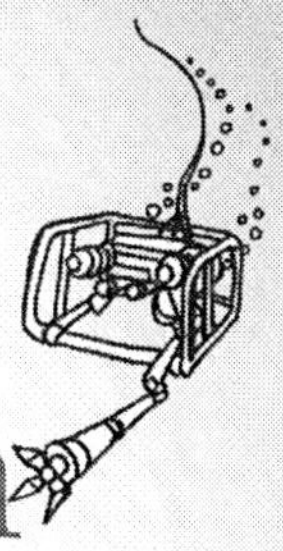

Rachel McDonald
Tina Miller-Way
Dauphin Island Sea Lab

Dauphin Island Sea Lab
Discovery Hall Programs
Education and Outreach

Who we are…

Dauphin Island Sea Lab Alabama

- Marine research and education facility for the state of Alabama
- Education program – Discovery Hall
  - o Field trip facility (with room & board) during the school year
  - o Summer camps – day and overnight; all ages
  - o Residential class in marine science for high school students
  - o Professional learning workshops for teachers (most free of charge)
  - o Traveling marine science 'lab' - the BayMobile; visits schools and festivals
  - o Public aquarium – *The Estuarium*

In with the new

- New standards - disciplinary core ideas
- Incorporation of science & engineering practices
- Focus on STEM (STEAM) activities
- Addition of engineering, robotics

- Bringing ROVs into your classroom, club or group – a way to bring these requirements altogether in one activity

The ROV umbrella. . .

- Physical science
  - o Forces, density, buoyancy, hydrodynamics, circuitry
- Biology
  - o ocean literacy, adaptations & biomimicry
- Mathematics
  - o conceptual, practical
- Even history
  - o exploration, technology

The ROV umbrella…

- Engineering
  - o Engineering design process
    - What do we need to design to achieve the mission?
    - Plan, build, test, assess, improve
    - Critical thinking

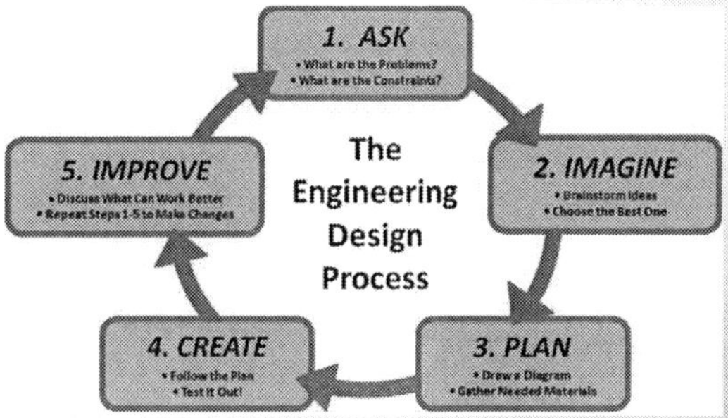

Other benefits

- Teamwork skills
- Career exploration –
  - o academic or career tech approach
- Additional complexity – design and skills possible

MATE program student survey results: 2015

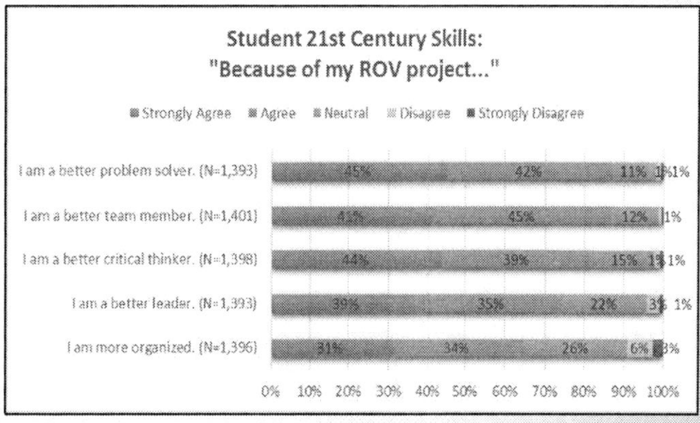

And bring in the ocean…

2013 (DISL):
Tropicalization of species

•Locate, identify, census, collect species

2014 (MA

•Locate, i ies

2014 (DISL), 2015 (MATE):
Oil Fingerprinting

•Collect and identify oil sample (based on fingerprint)
•Determine if from Macondo well

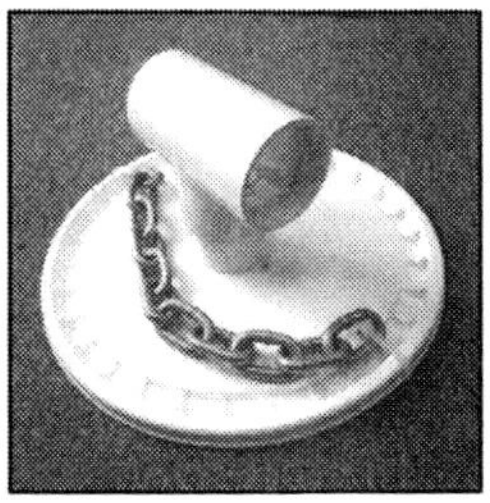

ROVing the Gulf:

Class components

$7^{th} - 12^{th}$

- History of deep sea exploration
- Hands-on activities- buoyancy, density
- Designing, building, redesigning a ROV (engineering design process)
- Hydraulic arm activity
- Assessing, evaluating…

Expanding our ROV offerings

- ROVing the Gulf (introduced 2011)
- Teacher workshop : *Technology in Marine Science* (introduced 2012)
- Student ROV competition (introduced 2013)
- Marine DeTECHtives (introduced 2017)
- Activities in summer camps
- ROV Outreach

Technology in Marine Science

- Annually since 2012
- Hands-on approach
  - Research vessel cruise
    - Drive scientific ROV
    - Use common technologies (CTD) & sampling gear
  - Classroom
    - Build ROV from kit (wiring, frame, control box, motors)
    - Collect & interpret CTD data
    - Hydraulic arm, build a sensor

Student ROV Competition

- Annual – in our 6th year; MS-HS

ROV Comp #s by Year	2013	2014	2015	2016	2017	2018
# of teams	6	6	12	16	23	TBD
# students	29	32	64	85	163	TBD

MATE

- Design and build ROV for missions/tasks
  - Opportunities for increased complexity (coding, hydraulics, etc)
- Engineering presentations
  - Local judges from engineering firms, universities, tech jobs, etc
- Marketing poster

2016 NGC Regional

Marine DeTECHtives

- STEAM based 3-day
  - Technology in a bag – wha
  - Design a vessel
  - make a mini robot
  - 1 hour of code (https://cod
  - ROVs – design, build, test,
  - vessel trip – trawl to collec
  - circuits & circuit art – mak boar

ROV Outreach

Discovery Hall Programs
Education and Outreach

- Middle school classroom visits
  - Maker Day
  - ROV 101 – background, drive ROVs, competition
  - STEM Days
- High school classroom visits
  - ROV 101 – background, drive ROVs, competition
  - STEM Days
- Regional and State competitions
  - Jubilee BEST (Mobile Co) & South's BEST Regional
- Career Expo
  - Hands-on career expo for Lower Alabama 8th graders
- Alabama Math, Science, Technology Initiative (AMSTI) Festival (AL Department of Education)

Impacts – ROV class

2012-2017

Pre/post assessments
10 question quiz

	pre	post
Mean (		
SD		
N 2		

Significant increase in content knowledge (P<0.0001)
increase from pre- to post-test

grade
school type
gender

Impacts – ROV Competition

MATE

Post Competition Survey 2015

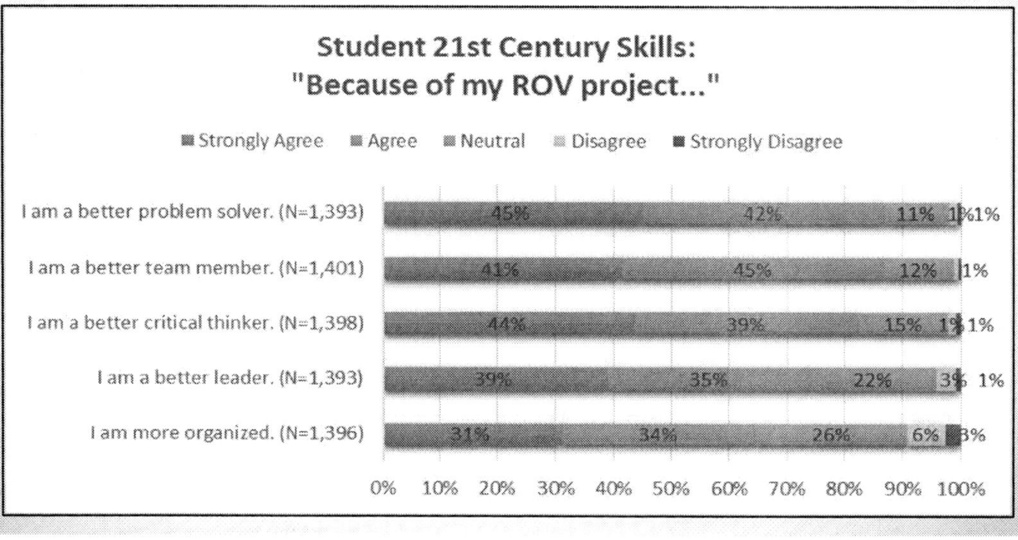

Impacts – Workshop

	Very Valuable	Valuable	Average
ROV building	100%		
Circuit diagram/ art	75	25	
Classroom Activities (buoyancy, physical principles)	75	25	
Guest speaker	92	8	
Vessel field trip	67	33	
Value of ROVs in STEM	67	33	
MATE competition	92		8

- What are the 3 most valuable components?
  - Building (6+), Circuitry, Resources, Hands-on experience, connections/discussions with colleagues, cross-curricular aspects

Impacts – TECH camp

- Reaching NEW campers
  - o Have you been to a camp at the Sea Lab before? (yes or no)
    - Yes - 30% **No - 70%**

- Tell us one thing you learned this week
  - o PCB abbreviation/ circuit board/ simple circuits: 25%
  - o ROVs – What are ROVs, How to build ROVs, ROVs are cool: 20%
  - o Bioluminescence 15%
  - o How to solder 5%
  - o Scratch and PCBs 5%
  - o Team work 5%

A big
THANK YOU
to
DHP Educators & Staff
Sponsors

Questions?

Dauphin Island Sea Lab
Discovery Hall Programs
Education and Outreach

http://www.disl.org/
http://www.disl.org/rov

Deep-C CONSORTIUM

GULF OF MEXICO RESEARCH INITIATIVE

MATE

GULF OF MEXICO ALLIANCE

MORRISON
CONTRACTORS
COMPANY OVERVIEW
Safety, Accountability, Quality, Sustainability in Offshore Operations
February 2018

LOUISIANA
HOUMA - CORPORATE HEADQUARTERS
9 Bayou Dularge Road
Houma, Lousiana 70363
HARVEY
3434 Peters Road
Harvey, Louisiana 70058
DEEPWATER RISER SERVICES
142 Industrial Blvd
Houma, Louisiana 70363
TEXAS
HOUSTON
16285 Park Ten Place,
Suite 100
Houston, Texas 77084
CARIBBEAN
WEST INDIES
Labidco Industrial State
7B St. New Jersey
La Brea Trinidad
West Indies
MEXICO
MEXICO HEADQUARTERS
Paseo Costa de Oro
648 Desp. 21-22 Boca Del Rio
Veracruz CP 94299
ALVARADO
Prol. Juan Soto S/N, Col. La Fuente.
CP Alvarado
Veracruz 95250
CD. del CARMEN
Calle 35 C No. 33
Col. Malibrán CP. Cd. del Carmen
Campeche 24118
MORRISON
CONTRACTORS
WWW.CHETMORRISON.COM

HSE Management System

This is not only way to successfully integrate HSE MS in business.

Rather this is our way. It is no perfect. How ever it gives our business case some competitive advantage and we are willing to share high points.

- *Risk Based approach*
  - *Leadership*
  - *Accountability*
  - *Risk Mitigation through implementation of most effective control*
- *Continual Improvement Approach*
  - *Plan*
  - *Do*
  - *Check*
  - *Adjust*

  - *ROI Based Approach*

"It is immutable law of business that words are words, explanations are explanations, promises are promises but only performance is reality."

HSE Management System

Goal Setting and Benchmarking
Implementation
Effectiveness Monitoring
Adjustment

By executive management in real time

HSE Performance Task
Man-Hours YTD
TRIR
ZAPP Participation
Near Miss : Incident Reports
HIPO/Man-hour
Hazard Hunts
HSE Committee Meeting
Line Management Visits
HSE Progressive Audits
MOC/ MOC Exercise
HSE Management Performance Review
JSA Desktop Audit
HSE Competency Assessment
HSE Leadership Assessment
Cost of Claim/ Man-hour
Man hours/ Medical Visit
Cost of Equipment Damage/ Man-Hour
Spill Reduction Goal
[illegible]

HAZARD SEVERITY OUTCOME			PROBABILITY					
INJURY	SPILL / POLLUTION	DAMAGE	Very Unlikely	Unlikely	Possible	Likely	Very Likely	
Very Serious Death or Multiple Serious Injuries	> 100 Gallons	US$ >1 million	M	M	H	H	H	E
Serious Long term serious injuries	50–100 Gallons	US$ >200,000	M	M	M	H	H	D
Moderate Injury leading up to 10 DAFW	10-50 Gallons	US$ >50,000	L	M	M	M	H	C
Minor First Aid required or medical treatment	1-10 Gallons	US$ >10,000	L	L	M	M	M	B
Slight No specific treatment or loss of work	Any Amount -1 Gallon	<10,000	L	L	L	M	M	A

Knowledge Areas	Project Management Process Groups				
	Initiating	Planning	Executing	Monitoring & Controlling	Closing
Project Integration Management	4.1 Develop Project Charter	4.2 Develop Project Management Plan	4.3 Direct and Manage Project Work	4.4 Monitor and Control Project Work 4.5 Perform Integrated Change Control	4.6 Close Project or Phase
Project Scope Management		5.1 Plan Scope Management 5.2 Collect Requirements 5.3 Define Scope 5.4 Create WBS		5.5 Validate Scope 5.6 Control Scope	
Project Time Management		6.1 Plan Schedule Management 6.2 Define Activities 6.3 Sequence Activities 6.4 Estimate Activity Resources 6.5 Estimate Activity Durations 6.6 Develop Schedule		6.7 Control Schedule	
Project Cost Management		7.1 Plan Cost Management 7.2 Estimate Costs 7.3 Determine Budget		7.4 Control Costs	
Project Quality Management		8.1 Plan Quality Management	8.2 Perform Quality Assurance	8.3 Control Quality	
Project Human Resource Management		9.1 Plan Human Resource Management	9.2 Acquire Project Team 9.3 Develop Project Team 9.4 Manage Project Team		
Project Communications Management		10.1 Plan Communications Management	10.2 Manage Communications	10.3 Control Communications	
Project Risk Management		11.1 Plan Risk Management 11.2 Identify Risks 11.3 Perform Qualitative Risk Analysis 11.4 Perform Quantitative Risk Analysis 11.5 Plan Risk Responses		11.6 Control Risks	
Project Procurement Management		12.1 Plan Procurement Management	12.2 Conduct Procurements	12.3 Control Procurements	12.4 Close Procurements
Project Stakeholder Management	13.1 Identify Stakeholders	13.2 Plan Stakeholder Management	13.3 Manage Stakeholder Engagement	13.4 Control Stakeholder Engagement	

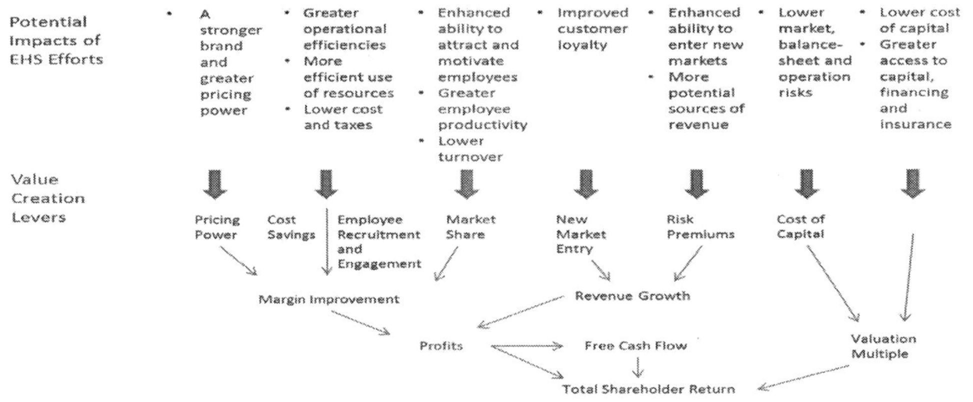

HSE not revenue generating center. But it is run like one. Rather revenue saving center.

HSEQ MS Performance

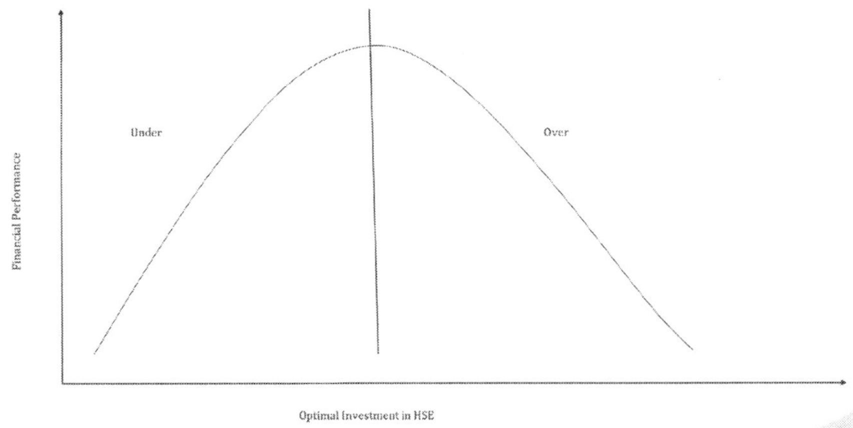

Accountability — Evidence, not emotion

↓↑ There is less to fear from outside than from inside - inefficiency, discourse and poor service

Quality MS

When people and organizations focus primarily on quality, defined by the following ration:

$$\text{Quality} = \frac{\text{Results of work efforts}}{\text{Total Cost}}$$

Quality tends to increase and costs fall over time

However, when people and organizations focus primarily on *costs,* costs tend to rise and quality declines over time

Human Element Current:

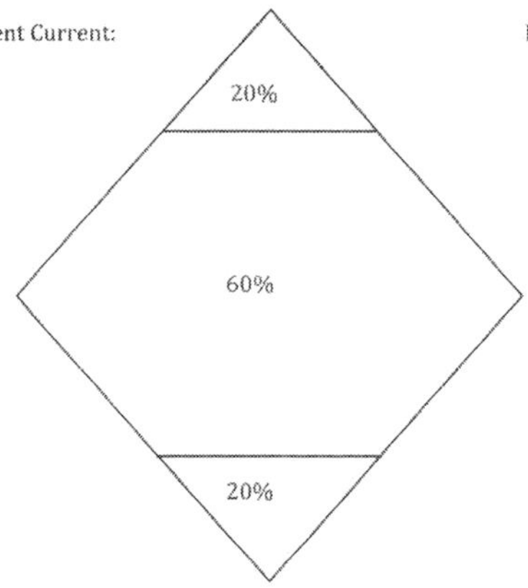

Human Element Desired:

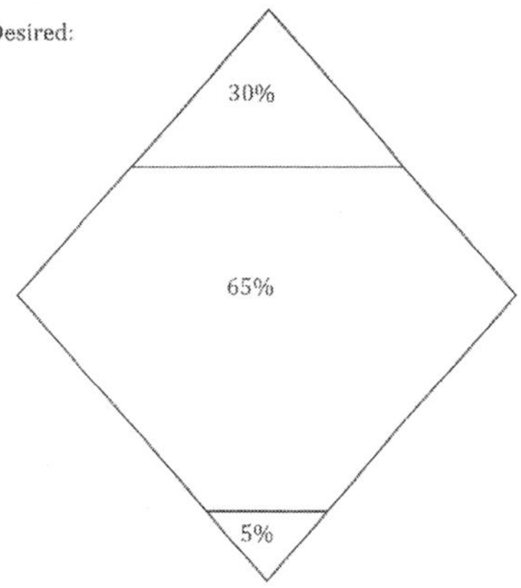

Training/Human Element Development

- Focused approach on key functions development
- Desired ROI% = $\frac{\text{Monetary Benefits} - \text{Training Cost}}{\text{Cost}} * 100$

You'll need to track the cost of your training program in order to complete this formula. Make sure that you know how much you've spent on:

- Design and development
- Promotion
- Administration
- Delivery (staff or technology)
- Materials
- Facilities
- Employee wages
- Evaluation

To track the monetary benefit of your monetary benefits of your training, you'll want to keep track of:

- Labor savings
- Productivity increases
- Income generation
- New leads
- New products
- Lower turnover costs

FFD & Work Readiness

- NWR - 16
- 20K man hours/doctor visit

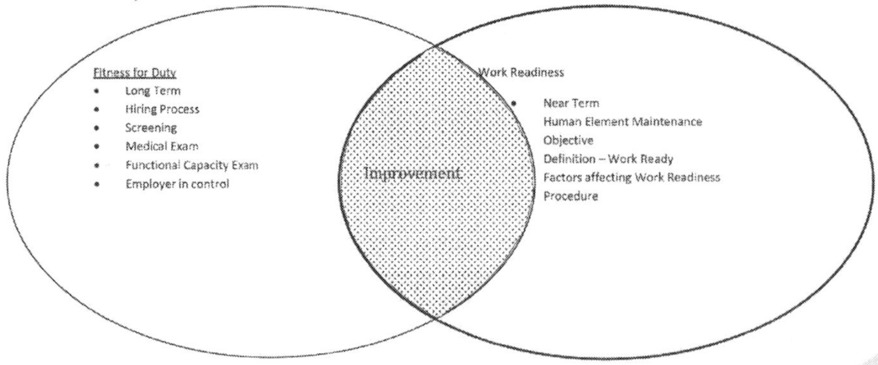

MORRISON
CONTRACTORS

Questions?

Thank you

Integrated ROV Systems

How an Integrated State-of-the-Art ROV System was used in the Discovery of the USS Indianapolis

Underwater Intervention 2018

GREENSEA

Greensea Mission & Strategy

Greensea improves the relationship between man and machine to make the work they do together more productive.

MAN AND MACHINE ENGAGEMENT

OPENSEA OPERATING PLATFORM

MANNED - UNMANNED - SURFACE - SUBSEA

NAVIGATION | **CONTROL** | **USER INTERFACE**

VIA 9GAG.COM

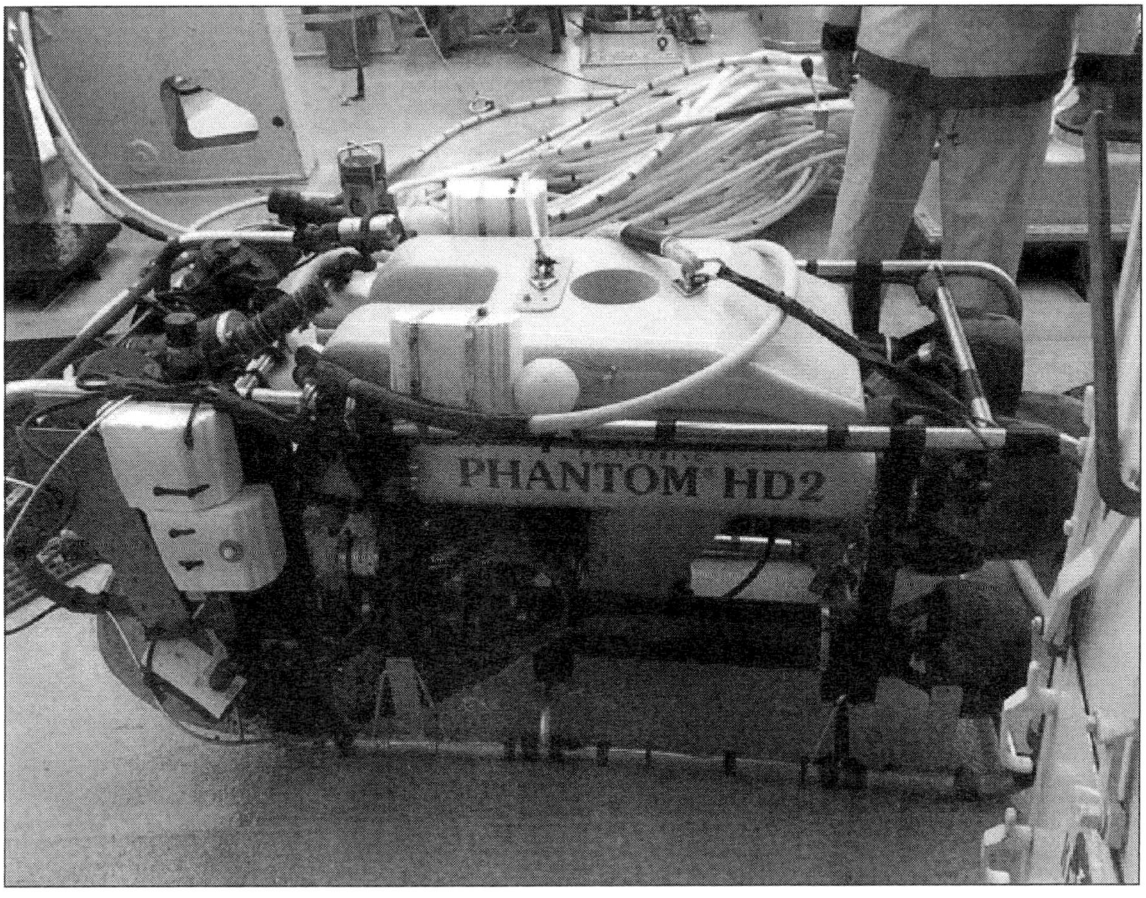
PHANTOM HD2

AT&T
2:17 PM
Done
Postrio Cafe & Bar
6:45PM 7:00PM 7:15PM
Post St
Geary St
Taylor St
Curran Theatre
Mason St
O'Farrell St
Cosmo Pl
Refresh
AT&T
2:18 PM
Back
Millennium
Details
Diner Reviews
Fri. Jul 13, 7:00PM for 2
6:45PM
7:00PM
7:15PM
Downtown / Union Square
Vegetarian / Vegan
$30 and under
580 Geary St.
San Francisco, CA 94102
Menu
Millennium writes:
OpenTable

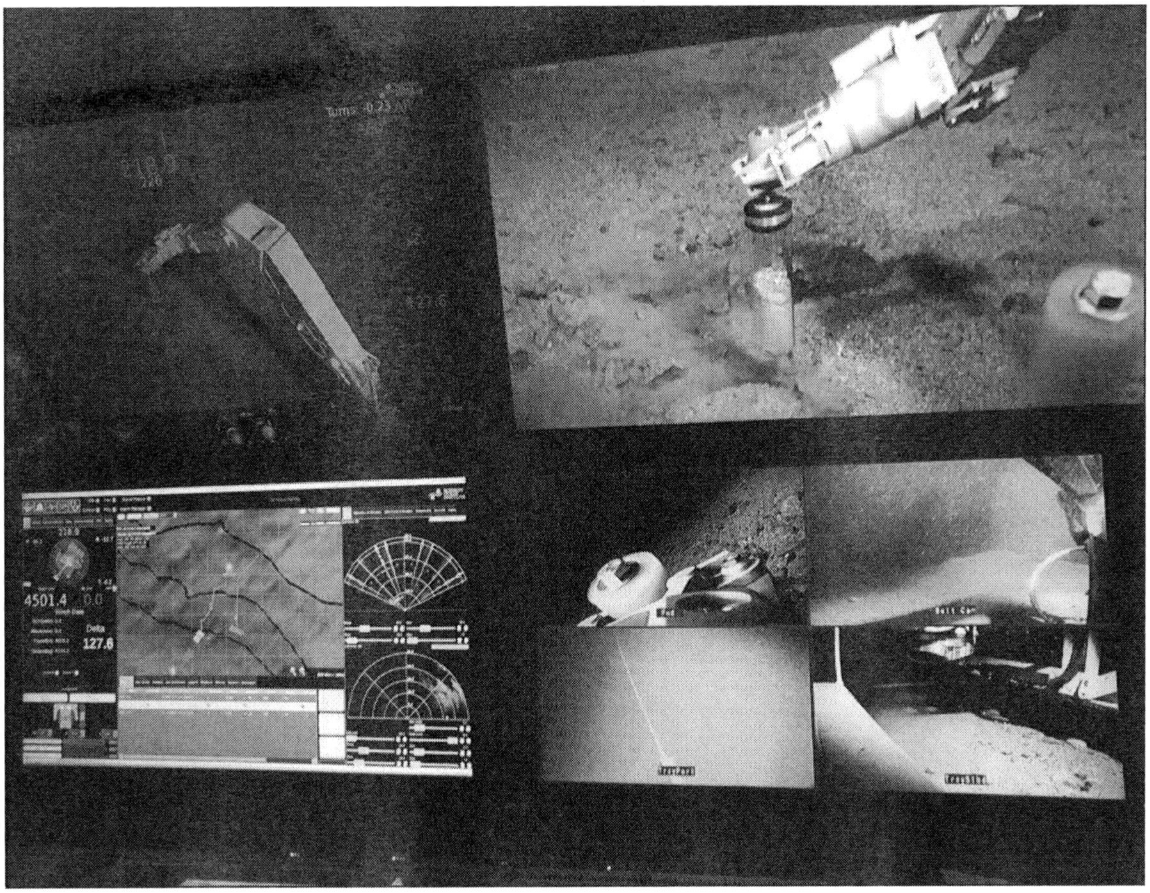

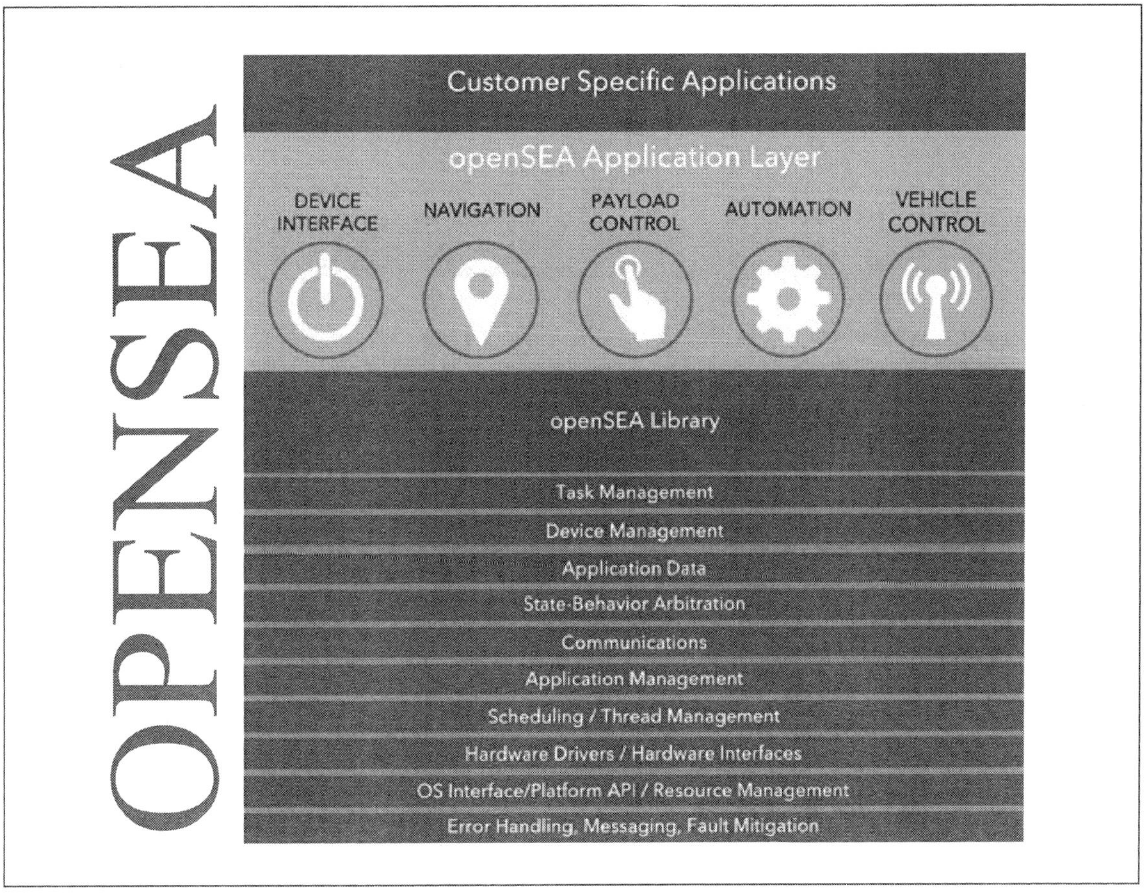
OPENSEA
Customer Specific Applications
openSEA Application Layer
DEVICE INTERFACE
NAVIGATION
PAYLOAD CONTROL
AUTOMATION
VEHICLE CONTROL
openSEA Library
Task Management
Device Management
Application Data
State-Behavior Arbitration
Communications
Application Management
Scheduling / Thread Management
Hardware Drivers / Hardware Interfaces
OS Interface/Platform API / Resource Management
Error Handling, Messaging, Fault Mitigation

With Integration, Vehicles Become Systems of Systems

Inspection-Class ROVs

Work-Class ROVs

Diver Propulsion Sys

Vessels

Sensors & Bouys

USVs

Towfish

Submarines

Integrated Sensors/Equipment Provide Integrated Data, Elevating Operator Awareness

Sonar

Doppler Velocity Log

USBL

Light & Camera

CTD, Pressure

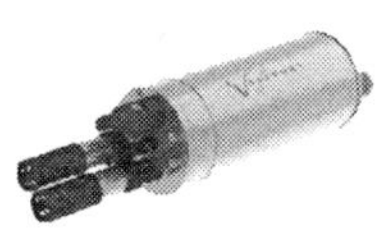

Operator Chair

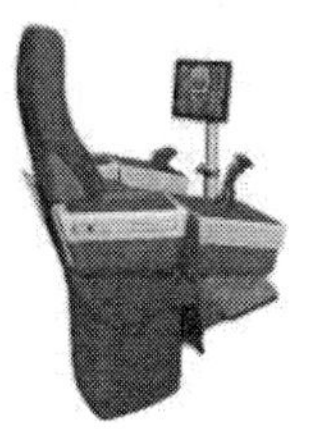

Joystick, throttle

Winch

10

Integration Allows us to Reframe the Conversation

Who's responsible for performance?

A Unified Operator Visualization Tool Organizes All System Data

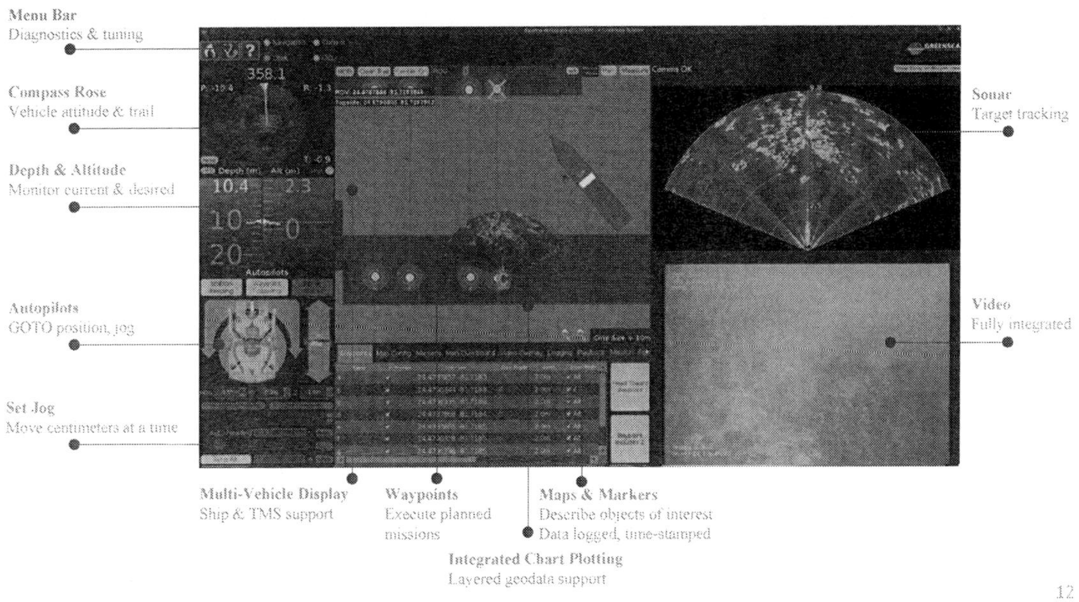

12

USS INDIANAPOLIS
VULCAN INC 6000M ROV
POWERED BY GREENSEA

15

- **Hydroid REMUS 6000**
- **Edgetech 2205 75kHz/230kHz Side Scan**

16

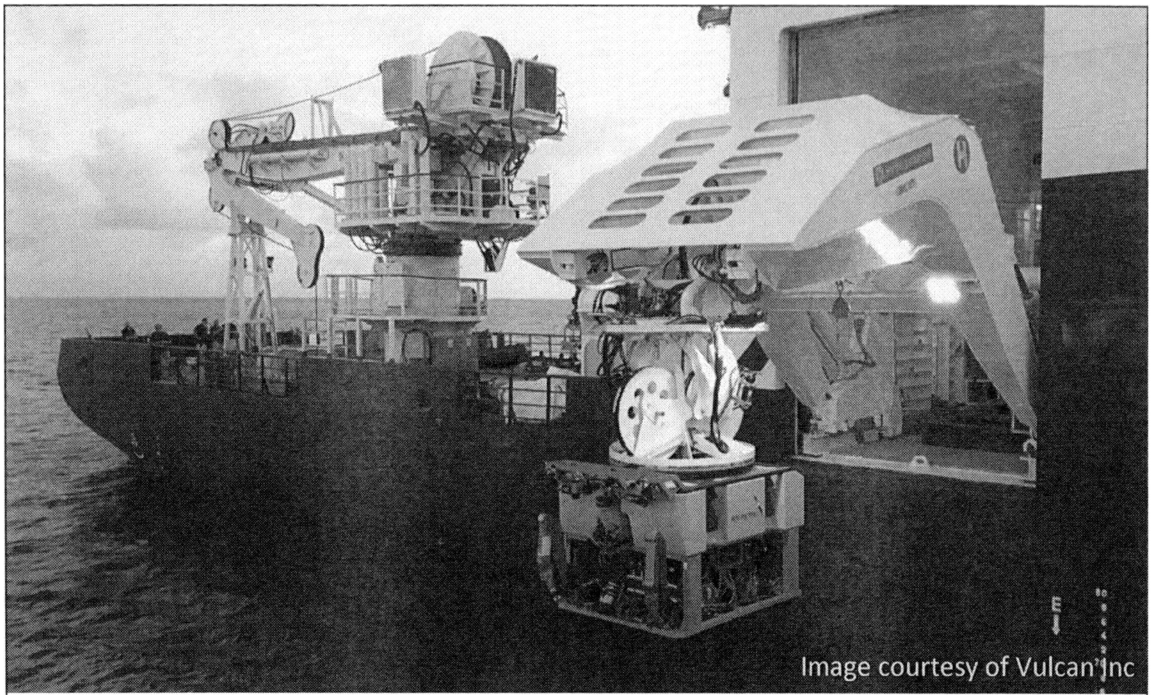

- **ARGUS BXL79 6000m ROV**
- **Greensea Inertial Navigation System**
- **Greensea Systems Control & Autonomy**

GREENSEA 17

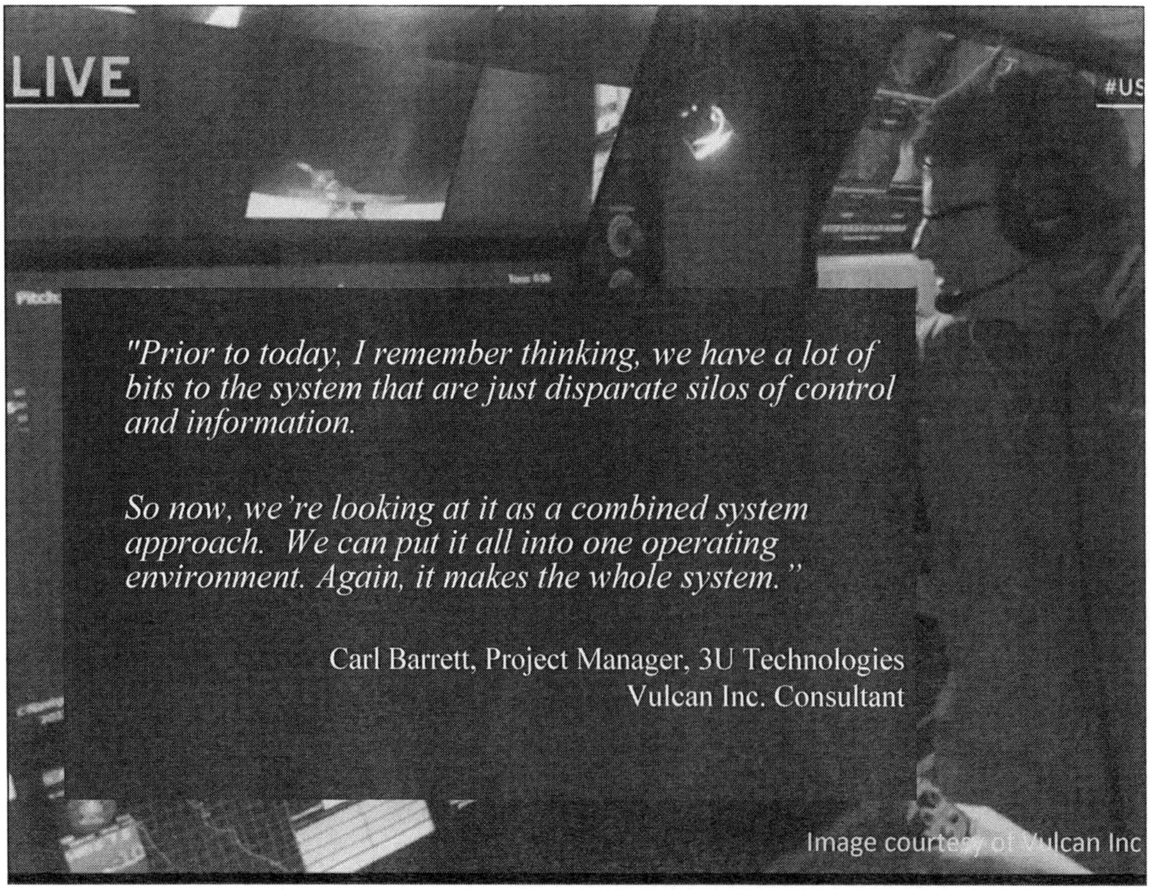

GREENSEA 19

Integrated ROV Technologies

Integrated technologies helped create operational efficiencies for the pilots and streamlined the mission's operations:

- **Automated ascent and descent winch control.**
  - Allows the pilots to set a desired vehicle depth and automatically descend or ascend to that level.
- **Automated black box data recorder.**
  - Turned on and off when the vehicle entered and exited the water, providing a comprehensive recording of the mission's telemetry data, for both the AUV and the ROV.
- **Synchronized pilot and co-pilot chairs with touchscreen displays.**
  - Supported simultaneous operation of lights, cameras, and hydraulics. while enabling vehicle control to be switched between pilots.
- **Multi-vehicle user interface.**
  - Supported visualization and integrated data management.

GREENSEA 20

Vulcan Inc. Subsea Survey Process

- The AUV deployed with pre-programed search parameters for up to 20 hrs. This search / survey technique commonly referred to as "mowing the lawn".
- Once a search grid is complete, the AUV returns to a pre-programmed location where it is retrieved and the team downloads its data.
- While reviewing the data, the team looks for anomalies that indicate man-made objects or features that appear abstract to the surrounding geology.
- Then the targets or features of interest are subsequently explored with the ROV for positive identification. The ROV descends and ascends at 35 meters per minute, depending on depth; it can take up to three hours for the ROV to reach its target.
- The AUV data provides precise location information that allows the ROV to pinpoint previously detected targets. In Petrel's survey suite, the team reviews the video feed to determine what the AUV scan detected.

GREENSEA 21

GREENSEA 22

Up Next, Vulcan Partners with US Navy to Survey Aircraft in Philippine Sea

23

Thank you.

Marybeth Gilliam
Chief Operating Officer

www.greensea.com
802.434.6080

GREENSEA

Connectivity and ROVs: Going Beyond the Dive

Presented by: Ted Curley, CDO

aquabotix.com

About Us

- Founded Aquabotix in 2011 by Durval Tavares, CTO.
- Aquabotix Technology Corporation is a technology company that specializes in the design and manufacturing of underwater vehicles, underwater camera systems and cloud based storage.
- Aquabotix products are currently deployed in over 45 countries. Over 350+ Vehicles in the Market and 500+ Camera Systems.
- April, 2017: UUV Aquabotix Ltd (Aquabotix) announced that it was undertaking an initial public offering (IPO) on the Australian Securities Exchange (ASX).
- September 2017: New state-of-the-art 7,000 ft² facility in Fall River, MA.
- December, 2017: Hired new CEO, David Battista
- December 2017: Introduce new Hybrid Vehicle.

2 aquabotix.com

What does Aquabotix Make?

AquaLens Connect
Underwater Camera System

Endura
Inspection Class ROV

Integra
Hybrid AUV/ROV

4 aquabotix.com

ENDURA Remotely Operated Vehicle (ROV)

Features:
- •Five Torque Motors
- •Standard Sensor Package:
  - o Depth
  - o Temperature
  - o Orientation
  - o Leak Detection
- •1080p True HD Camera with Pan & Tilt
- •Depth Rating up to 300 meters
- •Payload Capability up to Six Pounds
- •High Intensity LED Lighting (4400 Lumens)
- •Available in different body lengths (8", 16", 24")

aquabotix.com

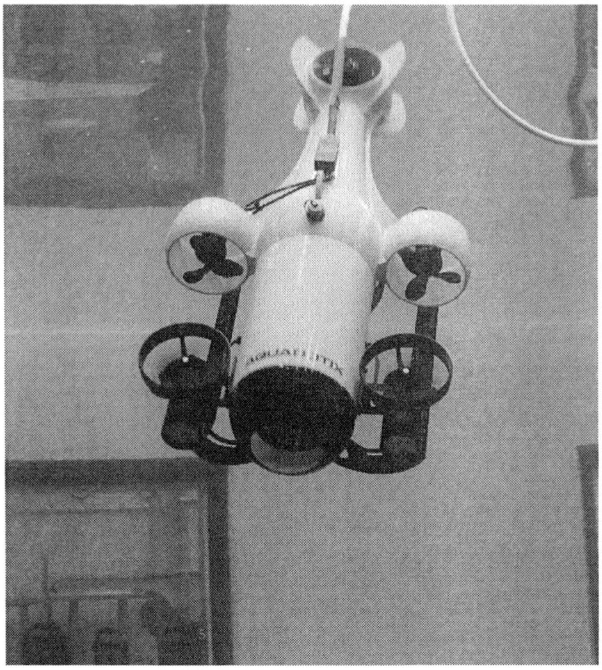

Sensor & Feature Options

- Side-Scan Sonar
- Scanning Sonar
- Multi-Beam Sonar
- Manipulator Arms
- Water Sampler
- Laser Scaler
- Thickness Gauge
- GPS
- Doppler Velocity Log
- USBL
- Environmental Sensors
- Tether Reel with Slip Rings
- First Person View Goggles
- User Control Workstation
- Fish Plow

6 aquabotix.com

INTEGRA Hybrid AUV/ROV
• Operation as either ROV or AUV force multiplier
• One Man Deployable
• Depth Rated up to 300m
• Run Time: 4+ hrs.
• Sensor Configuration:
o Side-Scan sonar
o INS
o DVL
o USBL
• Mission Planning: Intuitive Windows based program

Markets
Defense
Energy
Security & Law Enforcement
Research
Environmental Assessment
Aquaculture
Recreation
Infrastructure Assessment
8

Digital
Platforms and
Connectivity
aquabotix.com
9

Remote Viewing and Control
10

Air and Land Based Technology

Remote Live Viewing and Controls

Advantages

- Remote access from anywhere in the world
  - Especially helpful for companies with multiple offices
- Easy access from any web enabled device
- Operator does not need to be present
  - Reduction in cost of additional personnel/travel
- All sensor data available
- More than one user possible
  - Ability to set permissions, for example who can drive, who can only view, etc.

Digital Platforms for the Water World

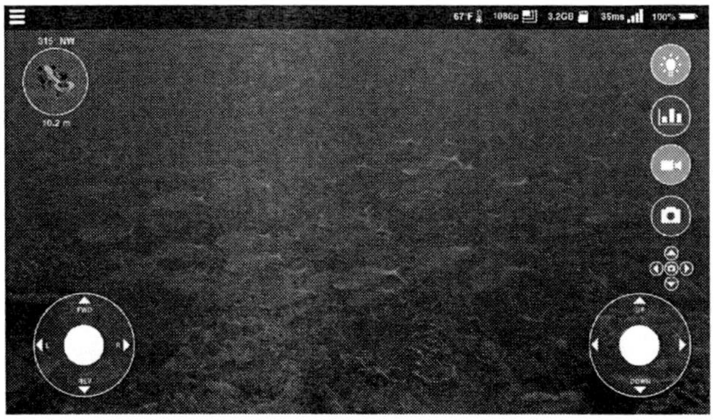

Fully Digital Platform using Web Based Technology

- Versus many who still have analog components
- Can be used on any laptop or iPad to control the vehicle, record photos/videos/data, and perform diagnostics
- Web Based User Interface is intuitive and easy to use and commands are performed at the touch of a button
- Parameters:
  - Internet connection required
  - 4G Hotspot
  - 5 MB/Sec. bandwidth
  - Good Ethernet Signal

Cloud Based Storage

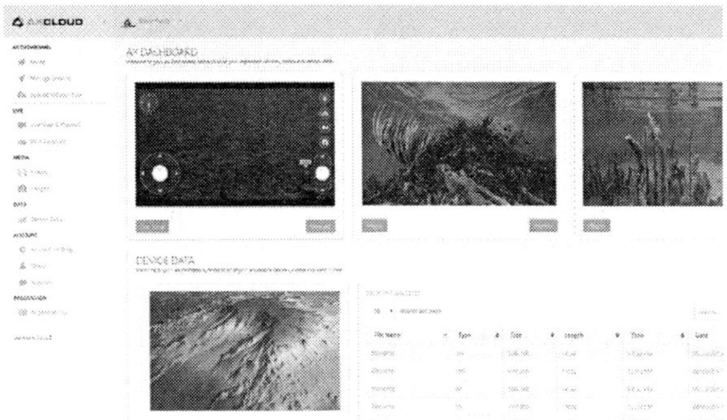

Advantages

- Ability to store and manage data from multiple assets (vehicles, cameras, sensors, etc.)
- Data stored all in one central place
- High Availability, High Security
  - Customer owns all their own data
- Files automatically synced via the cloud

Questions?

Supervised Autonomy: Humans AND Machines

Matt Ondler

HOUSTON MECHATRONICS
KNOW WHAT'S POSSIBLE.

Presenter – Matt Ondler

- President / CEO of Houston Mechatronics Inc.
- Formerly Aerospace executive
- 28 yrs at NASA, the last five as Chief of the Robotics and Software Division
- BS Aerospace Engineering, MBA Finance

HOUSTON MECHATRONICS
KNOW WHAT'S POSSIBLE.

HMI Overview

Former NASA Roboticists → Applying Technology to Industry

HOUSTON MECHATRONICS
KNOW WHAT'S POSSIBLE.

Houston Mechatronics Proprietary and Confidential

Highly Repeatable The Industrial Robot

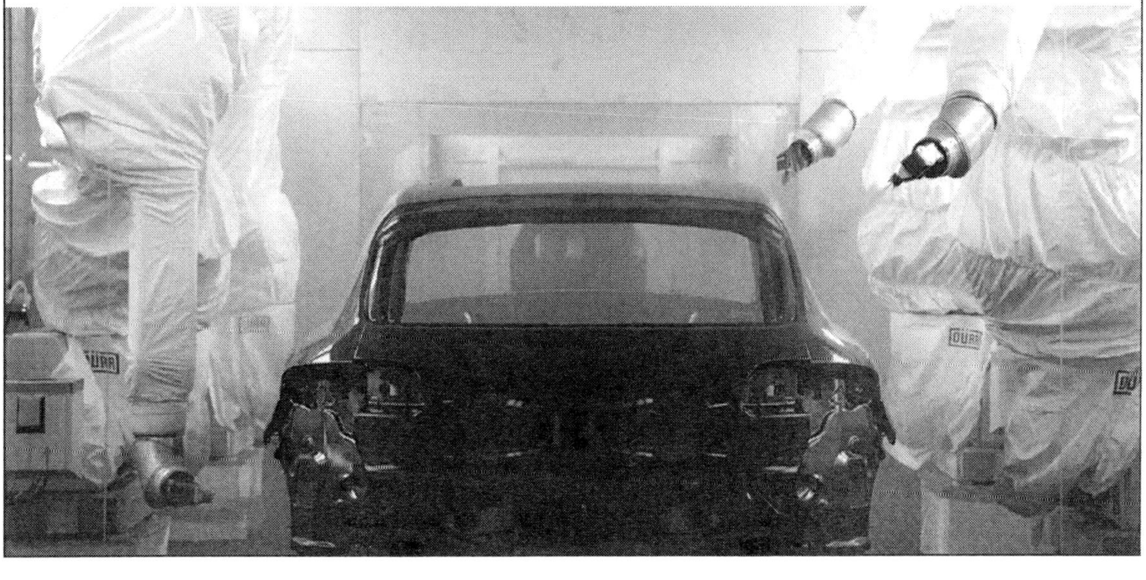

The Service Robot

Generally Repeatable
but Specifically Variable

Our World

Robot Expectations are High

But the reality is completely autonomous systems are a long way off…

…and turns out humans are really good at some things…

HOUSTON MECHATRONICS
KNOW WHAT'S POSSIBLE.

We Want Humans AND Robots

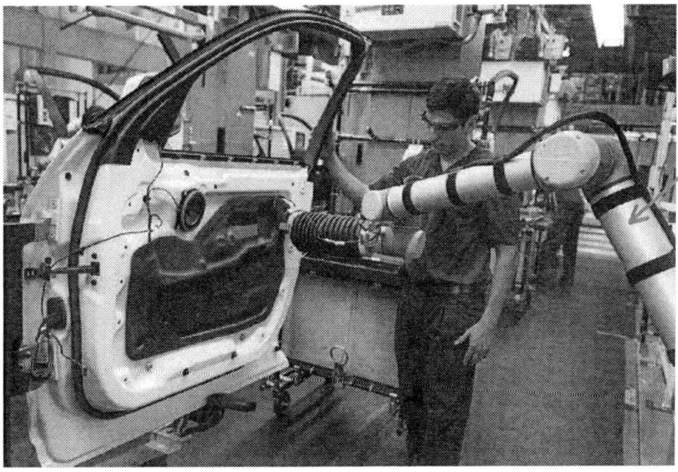

"Collaborative" Robot

Supervised Autonomy

We Want Humans AND Robots

Complimentary Advantages

Humans

- Mission planning and re-planning
- Safety and abort overrides
- Perception – especially in austere data environments
- High level problem solving
- Anomaly resolution

Robots

- Closed loop control
- Hazardous environments
- Repetitive precision
- High accuracy
- Number crunching
- Physical prowess (speed, strength, etc.)

Supervised Autonomy Use Case

Subsea ROV Operations

•Subsea ROV Operations require a team of people to perform (usually seven for 24 hour ops)

•Companies that provide these services have very little technology differentiation (many use the same vehicles)

•Very little technology evolution over time

•Operations are designed to be operator intensive and minimally automated (rationale is safety, but historically little pressure to automate)

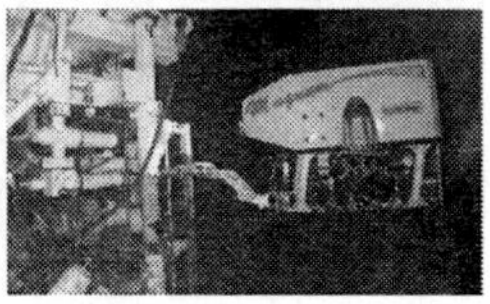

Supervised Autonomy Use Case

HMI's Command and Control Software

•Enables supervised autonomy (even under austere communication environments)

•Through data fusion creates immersive 3D environment for the operator

- Enhanced situational awareness

•Creates "affordance templates"

- Robot "understands" how to perform basic functions

•Control is abstracted to a higher level

•Can easily link together tasks to create missions

•Flexible, custom, and "right" levels of autonomy

•Maximizes effectiveness and efficiency of human AND robot

Design Goals

The foreseeable future is humans AND robots, but we want…

•Increased abstraction of control

- From joystick to mouse clicks

•Change the correspondence

- From *many people/one robot* to *many robots/one person*

•Leverage the best of human/machine capabilities

•Ability to customize level of supervision for the situation

•More intuitive human/machine interfaces

•Increased human situational awareness

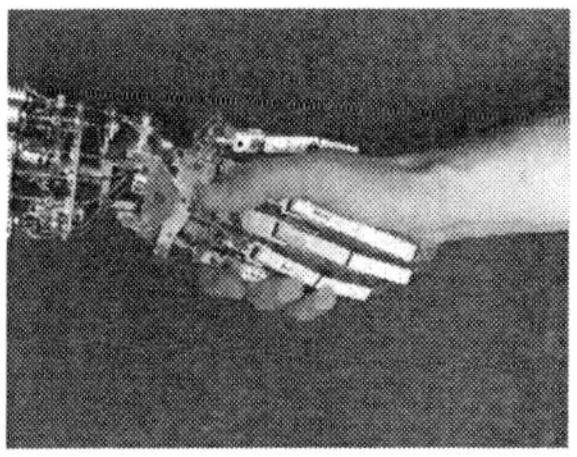

Questions?

Matt Ondler

HOUSTON MECHATRONICS
KNOW WHAT'S POSSIBLE.

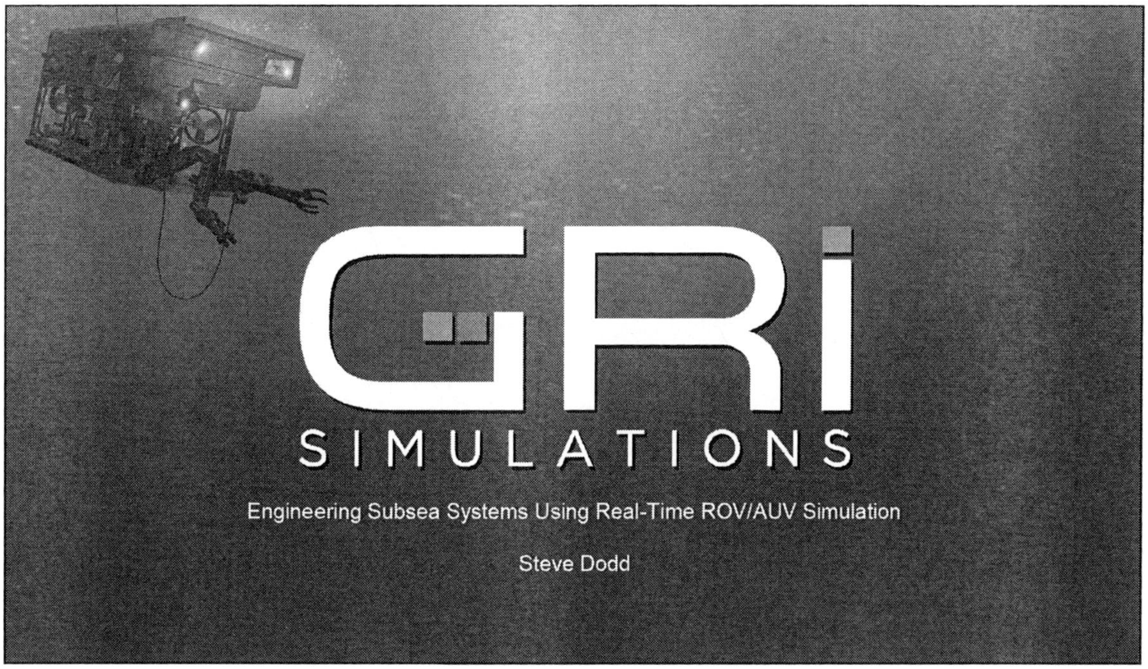
GRi
SIMULATIONS
Engineering Subsea Systems Using Real-Time ROV/AUV Simulation
Steve Dodd

GRi
SIMULATIONS
MTS
marine technology society
• Awarded the Compass Industrial Award 2017 by Marine Technology Society
• "For outstanding contributions to the advancement of the science and engineering of oceanography and marine technology"
GRi
SIMULATIONS
grisim.com

GRi
SIMULATIONS

- GRi Simulations and VROV
- Projects
  1) Sampler
  2) AIV-Autonomous Inspection Vehicle- i-Tech Services and Seebyte
  3) ROV3D Whitecap
  4) SLAM Sensor

GRi SIMULATIONS grisim.com

VROV Project Histories: Safe Marine Transfer

GRi SIMULATIONS grisim.com

VROV Project Histories:
Chevron PRS
73.7 MPa
55.6 MPa
36.5 MPa
86.9 MPa
128.0 MPa
106.6 MPa
61.4 MPa
75.5 MPa
102.8 MPa
99.7 MPa
GRI
SIMULATIONS
grisim.com

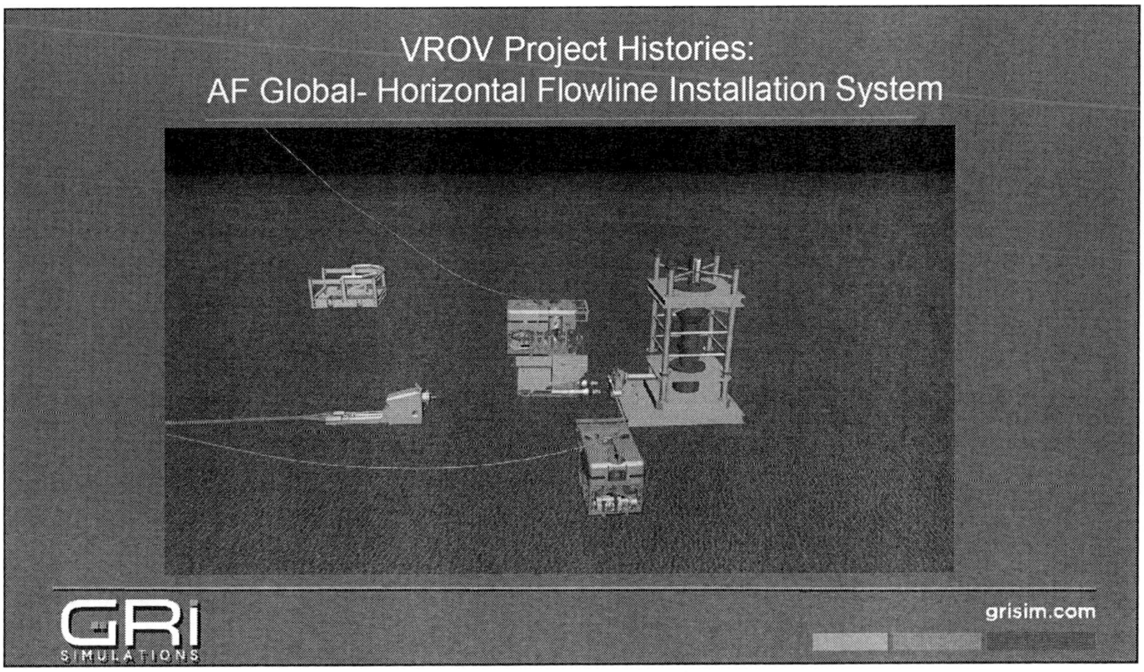
VROV Project Histories:
AF Global- Horizontal Flowline Installation System
GRI
SIMULATIONS
grisim.com

VROV Project Histories:
AF Global- Horizontal Flowline Installation System

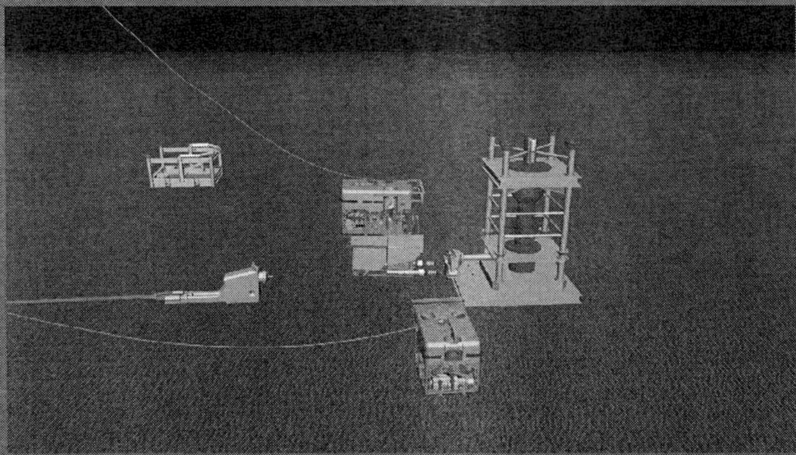

GRi SIMULATIONS

grisim.com

VROV Project Histories:
Trendsetter- E-SIT

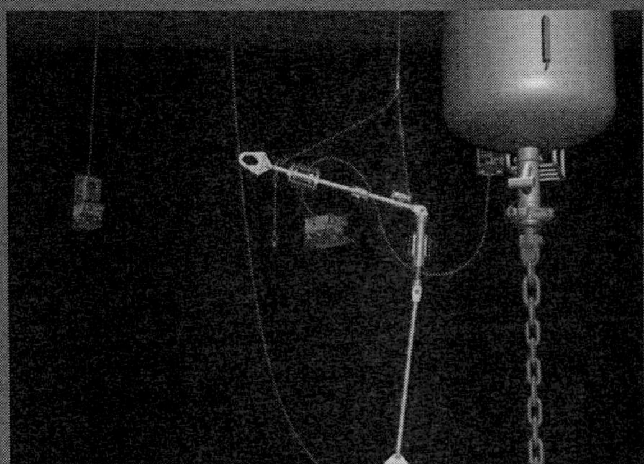

GRi SIMULATIONS

grisim.com

2) AIV-Autonomous Inspection Vehicle- i-Tech Services and Seebyte

GRI SIMULATIONS

grisim.com

Concept: In-Field Resident Autonomous Inspection and Intervention System

Value Proposition:

- True goal based autonomy and reduced vessel dependency
- Reduced Costs for Routine Inspection Option
- Continuous Availability instead of Seasonal Inspection Program
- Enabler for continuous monitoring, big data and digital transformation

GRI SIMULATIONS

grisim.com

Engineering Support: Subsea7 AIV

grisim.com

AIV Simulator Testbed

grisim.com

Role of 3D ROV Simulation:

Fully integrated Virtual AUV and Environment testbed. Unplug the ROV joysticks and screens and plug in an AIV brain, for the development and testing of:

- Control Software and Sensor Interfaces
- Advanced Autonomy Algorithms
- AIV Hardware Design
- Soak testing of AIV computing platform, hardware-in-the-loop
- Dry system integration and test

GRI SIMULATIONS

grisim.com

Benefits of 3D Simulation:

- Significant reduction in Wet Testing Requirements– reduced cost
- 24hr, 7 days availability
- 3D View of AIV Behaviour During Simulated Operations
- Reduced Risk in Wet Trials
- Engineering Efficiency: Design/Test Iterations Time Greatly Reduced

GRI SIMULATIONS

grisim.com

Simulation Special Requirements:

- Hydrodynamic AUV Behaviour
- High reliability and repeatability
- High accuracy simulation of Sonar Images and Formats – noise and artefacts included
- 4 simultaneous sonar simulations – 3 x Blueview and 1 x Tritech
- Accurate camera field of view simulation
- Simulation of acoustic modem system
- Multiple navigation sensors simulation, including DVL behavior on structures
- Controls over sensor quality, water current and visibility.
- Map export of simulation scenarios (ESRI)

GRI SIMULATIONS

grisim.com

Results and Residual Benefits

- Ongoing System Development Testing and Verification:
- ROV Simulator as AIV System Operation Demonstration System
- Virtual Layouts for Mission Planning

GRI SIMULATIONS

grisim.com

3) ROV3D-- Whitecap: Enhanced Asset Integrity through Live 3D Reconstruction, Inspection, and Simulation

GRi SIMULATIONS

grisim.com

ROV3D Whitecap: Enhanced Asset Integrity through Live 3D Reconstruction, Inspection, and Simulation

- Integration of VROV and Whitecap Scientific's ROV3D.
- ROV3D uses video feeds to construct 3D models in real-time
- GRi and Whitecap have been working to implement these models into VROV for the purposes of simulation testing and analysis.
- Final result will be a product used for increased environmental and asset awareness in inspection tasks, particularly as they relate to unknown environments

GRi SIMULATIONS

grisim.com

Project Participants:
ExxonMobil
Husky Energy
SUNCOR
Statoil
Chevron
petroleum
RESEARCH
newfoundland & labrador
WHITECAP
SCIENTIFIC
CORPORATION
GRi
SIMULATIONS
GRi
SIMULATIONS
grisim.com

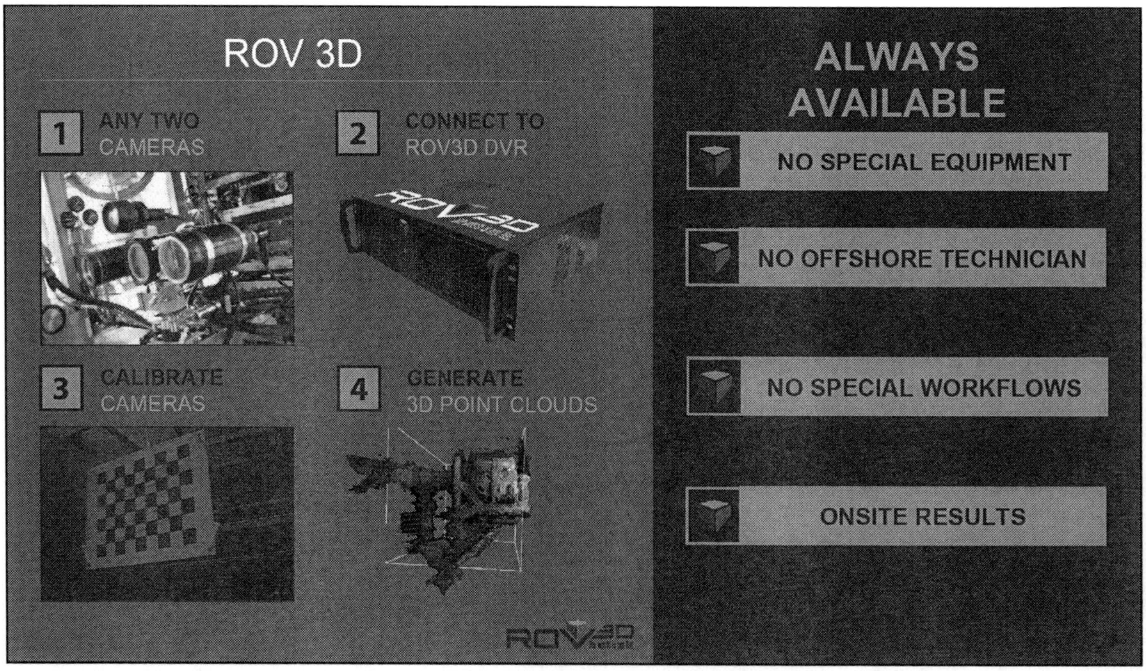
ROV 3D
1 ANY TWO CAMERAS
2 CONNECT TO ROV3D DVR
3 CALIBRATE CAMERAS
4 GENERATE 3D POINT CLOUDS
ALWAYS AVAILABLE
NO SPECIAL EQUIPMENT
NO OFFSHORE TECHNICIAN
NO SPECIAL WORKFLOWS
ONSITE RESULTS

Technical Objectives

Develop a prototype system supporting the following functionality:

- Collect metrically-correct 3D model data, whenever needed, using existing video inspection procedures, with the ability to make immediate measurements.
- Assess change relative to pre-existing CAD design, or previous inspection observation, in the field and in the office.
- Support import of 3D field model data into a simulator for further situational analysis and intervention planning, in the field and in the office.

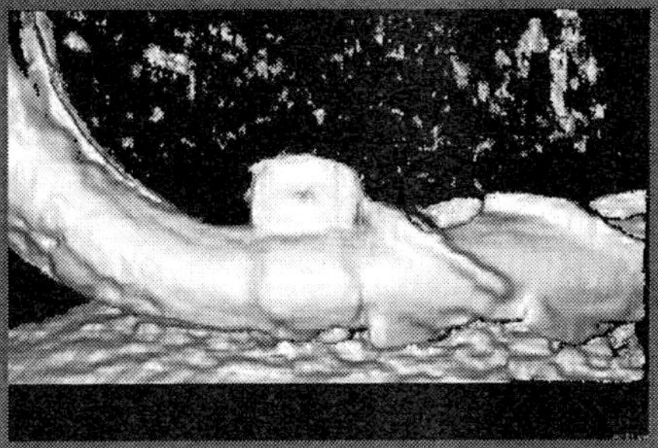

Project Milestones

- Does VROV/FDK remain responsive when working with high-polygon count meshes as generated by ROV3D? Establish mitigating measures as may be required
  - Mesh simplification and mesh decimation methods implemented
  - Texture handling strategy improved
  - Testing indicates good performance

Project Results

- Functional Integrated systems
- Established
  "Best-Practices" for 3D scene reconstruction
  - 1.5 – 2.5 metre standoff distance
  - <0.5-0.7 knot
  - <0.2-0.3 rad/sec (12-17 deg/sec) heading
  - Avoid glare (best to offset lighting)
- Change Detection Functionality

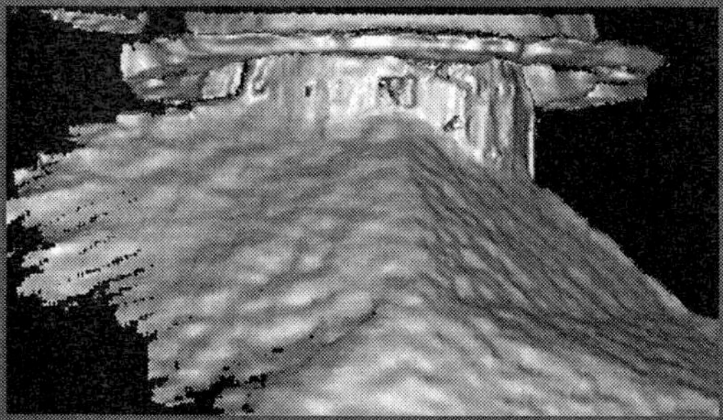

4) SLAM Sensor Development

GRI SIMULATIONS

grisim.com

Project: Integrated tank localization solution

Concept: Develop a smart sensor that provides simultaneous localization and mapping (SLAM) in feature-rich gps-denied environments such as tanks and ship hulls.

GRI SIMULATIONS

grisim.com

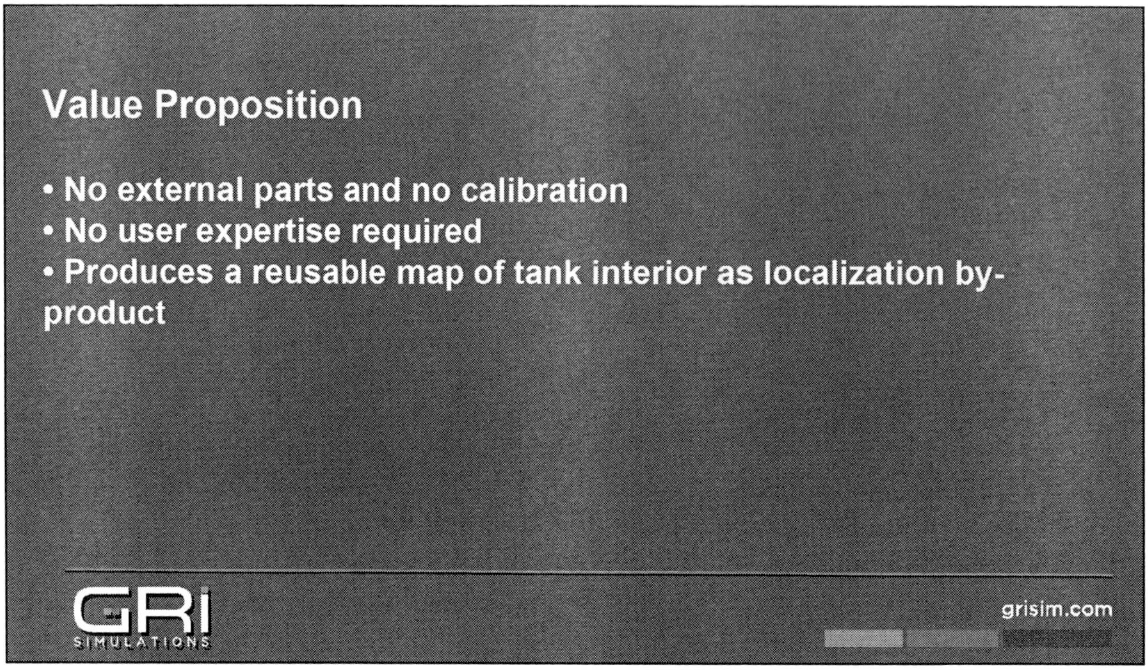
Value Proposition
• No external parts and no calibration
• No user expertise required
• Produces a reusable map of tank interior as localization by-product
GRI SIMULATIONS
grisim.com

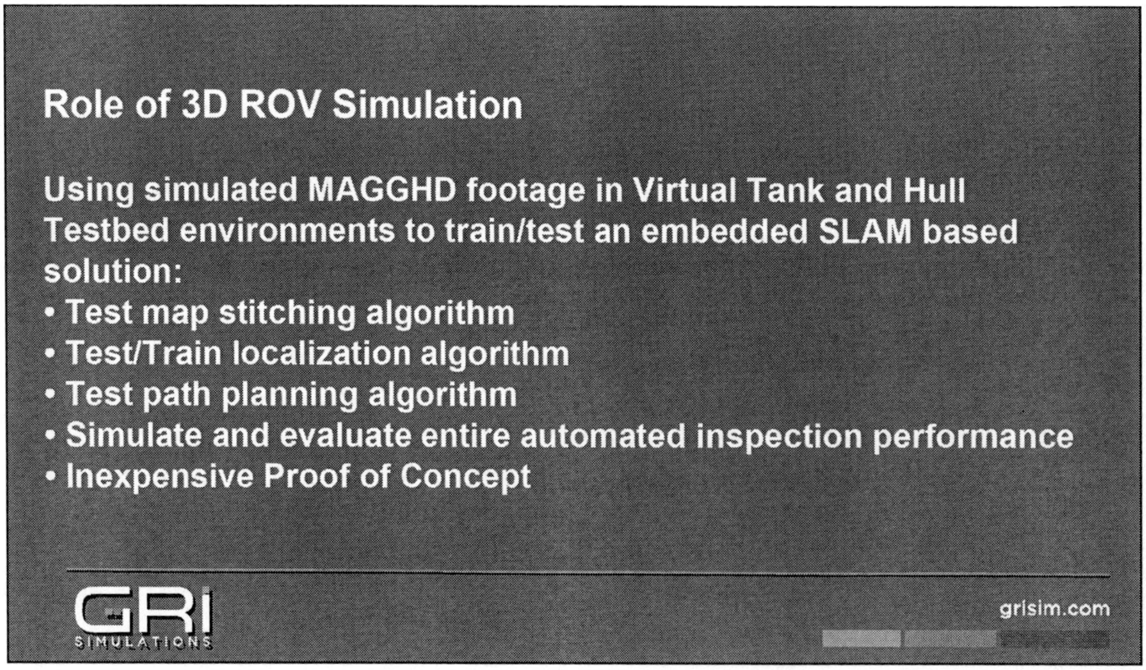
Role of 3D ROV Simulation
Using simulated MAGGHD footage in Virtual Tank and Hull Testbed environments to train/test an embedded SLAM based solution:
• Test map stitching algorithm
• Test/Train localization algorithm
• Test path planning algorithm
• Simulate and evaluate entire automated inspection performance
• Inexpensive Proof of Concept
GRI SIMULATIONS
grisim.com

Benefits of 3D Simulation

- Significant reduction in physical testing– reduced cost
- Ability to run continuously without human intervention
- 3D View of Platform Behaviour During Simulations
- Reduced Risk in Wet Trials
- Engineering Efficiency: Shortened Design/Test Iterations

GRI SIMULATIONS grisim.com

Simulation Special Requirements

- Hi-FidelityTracked ROV performance Behaviour
- Hi-Complexity Scenarios
- Hi-Fidelity simulation of high frequency multibeam sonar data and format
- Sonar must support ghosting, or reflection from environmental features

GRI SIMULATIONS grisim.com

Residual Benefits

- ROV Simulator as Sensor and Platform Operation Training System.
- ROV Simulator as Sensor Operation Demonstration System.
- Expandable to other similar applications like ship hull inspection

GRi SIMULATIONS

grisim.com

GRi

SIMULATIONS

Steve Dodd
Vice President, Operations & Business Development
steve.dodd@grisim.com
1-709-747-5599 (ext. 202)

GRi SIMULATIONS

grisim.com

AUTHOR INDEX

Anderson, B. 209
Baryal, N. 158
Bendzlowicz, M. 106
Bramlett, J. 186
Charan, S. 158
Covill, J. 263
Crouse, J. 177
Curley, T. 310
Detke, T. 87
Diaz, M. 158
Dodd, S. 325
Dunbar, T. 263
Gallagher, D. 79
Gardner, M. 229
Gilliam, M. 298
Hotaling, L. 51
Hughes, W. 79
Katt, D. 177
MacKay, I. 219
Manley, R. 79
McCarron, D. 19
Mcdonald, R. 281
McKone, K. 177
Merrill, B. 243
Miller-Way, T. 281
Ondler, M. 318
Pai, S. 38
Pilher, A. 79
Podrebarac, J. 263
Preston, J. 44
Ricci, W. 202
Ristic, B. 290
Rossland, R. 1
Simpson, D. 158
Smith, C. 177
Tena, I. 166
Zande, J. 99